Springer Series in Materials Science

Volume 236

The Springer Series in Materials Science covers the complete spectrum of materials physics, including fundamental principles, physical properties, materials theory and design. Recognizing the increasing importance of materials science in future device technologies, the book titles in this series reflect the state-of-the-art in understanding and controlling the structure and properties of all important classes of materials.

More information about this series at http://www.springer.com/series/856

Young Pak Lee · Joo Yull Rhee
Young Joon Yoo · Ki Won Kim

Metamaterials for Perfect Absorption

Springer

Young Pak Lee
Department of Physics
Hanyang University
Seoul
Republic of Korea

Joo Yull Rhee
Department of Physics
Sungkyunkwan University
Suwon
Republic of Korea

Young Joon Yoo
Department of Physics
Hanyang University
Seoul
Republic of Korea

Ki Won Kim
Department of Information Display
Sunmoon University
Asan
Republic of Korea

ISSN 0933-033X ISSN 2196-2812 (electronic)
Springer Series in Materials Science
ISBN 978-981-10-9082-0 ISBN 978-981-10-0105-5 (eBook)
DOI 10.1007/978-981-10-0105-5

Printed on acid-free paper

This Springer imprint is published by SpringerNature
The registered company is Springer Science+Business Media Singapore Pte Ltd.

Preface

In *The Invisible Man* (1897), H.G. Wells introduced the idea, then fictional, of a scientific route to invisibility. The protagonist used bleach and mysterious rays to make himself invisible.

This might be a time to write a book on metamaterial (MM) absorbers, since these days there are huge number of papers on this topic. In this way, one can accelerate something about the publicity situation of the field, which leaves something to be desired. At the same time, whenever a starting Ph.D. student wishes to work in MM absorbers, he or she has to fumble his/her way forward amidst a mass of scattered papers, unpublished notes, and Ph.D. theses, which are not all electronically available. It seemed to me that one coherent presentation might help.

The main purpose of this book is to provide consolidated information on the MM perfect absorbers. A brief history of the MM perfect absorber is incorporated to present the milestones of advancement. The theoretical backgrounds for and the fundamental of MM perfect absorbers are included to provide an insight on how the MM perfect absorbers work and we make them. It is also elucidated that, according to the operation frequency of electromagnetic wave on the MM perfect absorber, how differently the MM perfect absorbers are designed and made, and the behavior is changed. Methods to fabricate and characterize the MM perfect absorbers are presented in this book. The book emphasizes and elaborates the performance and characteristics of the MM perfect absorbers fundamentally, including the utilization of electromagnetically induced transparency, and practically. This book also provides recent advances on the MM perfect absorbers and the application aspects of MM perfect absorbers, such as multi-band, broadband, tunability, polarization independence, and incidence independence, to answer the present needs of the society. So far, few books have been published that incorporated the theoretical backgrounds up to the perspectives of MM perfect absorber, mainly covering many kinds of MM perfect absorbers, reflecting the practical aspects.

Few books are published in the field of MM perfect absorbers. This book includes practical aspects of MM perfect absorbers such as broadband and tenability, provides useful insights on the aspects of practical properties and fabrication of MM perfect absorbers as well as the fundamental and application perspectives, and puts many kinds of MM perfect absorbers reflecting the practical aspects, even the MM perfect absorbers utilizing electromagnetically induced transparency, the design and fabrication, the characterization, and the perspectives into one package.

It is my hope that at least the first three and the last chapters of the book are written in a sufficiently leisurely textbook style for undergraduate and graduate students with some requisite preliminary knowledge to read it. In the remaining chapters, which are a bit of 'capita selecta,' the style becomes more succinct and the aim is rather to summarize results and give a guide to the literature. This book presents essential information not only for undergraduate and graduate students but also for the people in academe and industry in their quest for the commercial exploitation of potential MM perfect absorbers. On the other hand, the students are better to have at least some knowledge of electromagnetism, optics, and condensed matter Physics, etc., before reading it.

Since this book contains a large part of our works done in the last decade, I should thank the people who were influential to us during that period. Since majority of what I report in this book has been joint works with others, I thank, especially, my main coauthors, Joo Yull Rhee, Young Joon Yoo, and Ki Won Kim. It is clear from the text to what extent we are indebted to B.S. Tung, J.S. Hwang, Y.J. Kim, N.V. Dung, and B.X. Khuyen, who helped us with the arrangement of data, the respective proof, and the discussions. Last but not least, my gratitude goes out to Mrs. C. Zitter, Physical-science Editor of Springer at Dordrecht, and Ms. A. Kang, Associate Editor of Physical Science and Engineering of Springer-Korea at Seoul.

We hope this book will convey the excitement of metamaterial absorbers to the readers and stimulate interdisciplinary interactions among researchers, thus leading to explorations of new frontiers.

Seoul, Korea Young Pak Lee

Contents

1 Introduction 1
References 5

2 Theoretical Backgrounds 11
2.1 Interaction of Electromagnetic Waves with Matter 11
2.1.1 Boundary Conditions and Fresnel Equations 15
2.1.2 Dispersion Relations 17
2.2 Perfect Electromagnetic-Wave Absorbers 20
2.2.1 Broadband Perfect Absorbers 20
2.2.2 Resonant Perfect Absorbers 27
2.2.3 Metamaterial-Based Perfect Absorbers 31
2.3 Effective-Medium Approximation 35
2.4 Equivalent-Circuit Theory and Transmission-Line Theory 39
2.5 Numerical Techniques 45
2.5.1 Finite-Difference Time-Domain Method 45
2.5.2 Finite-Element Method. 48
2.5.3 Transfer-Matrix Method 49
References 52

3 MMPAs Operating in Different Frequency Ranges 55
3.1 MMPAs Operating in the GHz Range 55
3.2 MMPAs Operating in the THz Range. 69
3.3 MMPAs Operating in the Infrared and Optical Ranges 79
References 96

4 MMPA, Based on Electromagnetically-Induced Transparency 99
4.1 Introduction 99
4.2 Narrow Perfect Absorbers Based on Plasmonic Analog of Electromagnetically-Induced Absorption 102

4.3 Broadband Perfect Absorbers Based on Asymmetry Resonators 106
4.4 Summary 110
References 111

5 Broadband and Tunable MMPA 113
References 140

6 Polarization-Independent and Wide-Incident-Angle Metamaterial Perfect Absorber 143
6.1 Introduction 143
6.2 Dependence of Electromagnetic Polarization on Metamaterial Perfect Absorber 144
6.3 Wide-Incident-Angle Metamaterial Perfect Absorber 154
6.4 Summary 165
References 165

7 Perspectives 169
References 172

Index 175

Chapter 1
Introduction

Abstract Recently, vigorous researches have been performed in the area of metamaterials (MMs). One outstanding effect is perfect-absorption MMs or MM-based perfect absorbers, that is, blackbody MMs. MM absorber has been firstly proposed in 2008, which had advantage of small size and thin thickness compared with the conventional absorbers. Since then, a great number of optimized MM absorber have been proposed for different application areas. Obviously, the MM single-band high absorption is inapplicable in some areas. Therefore, the research on broadband or multi-band high-performance MM absorber is necessary. Electromagnetic (EM) waves are in various polarization states, and to enhance the absorption the MM absorber should be designed to absorb EM waves independently of the polarization. The MM absorbers to be more practical should have the capability to cover large angle of incidence of the EM wave. THz or high-frequency MMs have received much attention, since conventional and natural materials hardly response to THz EM waves. Thus far, though most MMs were fabricated on rigid substrates, there have been several studies on flexible MMs. To achieve the perfect absorption, the method utilizing electromagnetically-induced transparency has also been investigated. Recent researches on MM absorbers and radiators include design of MM-based lenses and antennas, fabrication and measurements of MM structures for antenna applications, design and measurements of MM absorbing materials and screens, industrial applications of MM absorbers and radiators, etc.

Nowadays, it is said that we are living in the world of information. This means that we should exchange a huge amount of information with each other. Therefore, technologies are required to fulfill the desire of human being. In the forefront of technology, it is always essential to develop material possessing improved or new and/or novel properties useful for practical applications. This kind of needs has resulted in vigorous researches in the area of metamaterials (MMs) [1]. The term 'metamaterial' is used by Walser in 2001 for the first time [2]. A sophisticated definition is given by DARPA (Defense Advanced Research Projects Agency), USA, as MMs are a new class of ordered composites that exhibit exceptional properties not readily observed in nature. These properties arise from qualitatively

Y.P. Lee et al., *Metamaterials for Perfect Absorption*,
Springer Series in Materials Science 236, DOI 10.1007/978-981-10-0105-5_1

new response functions that (i) are not observed in the constituent materials and (ii) result from the inclusion of artificially-fabricated, extrinsic, low-dimensional inhomogeneities [3].

Recently, people like to use a simpler definition; MMs are artificial media structured on a size scale smaller than the wavelength of external stimuli [4]. A detailed discussion of terminology is given in [5].

During the past decade, materials artificially engineered, the so-called MMs, which possess unnatural electrodynamic properties and effects, such as negative refractive index [6], inverse Doppler effect [7], superlensing [8], and electromagnetic (EM)-wave cloaking [9], have attained great achievements in photonic researches [10–13]. Together with developments of nanotechnology, MMs have not only produced the fascinating effects in a wide range of EM wave, but also been gradually applied to epoch-making improvements of microwave and photonic devices by exploiting the advanced phenomena [14]. The manipulation of effective parameters in man-made media increases diversification in the application of MMs [15]. One outstanding effect is perfect-absorption (PA) MMs or MM-based perfect absorbers (MMPAs), that is, blackbody MMs, which are useful to enhance the efficiency in capturing solar energy [16] and applied to plasmonic sensor [17], bolometer [18], wireless power transfer [19] and perfect light absorber have been rapidly developed [20–22]. The resonance that is established between inductive and capacitive portions of the circuit allows energy to be stored and subsequently dissipated via Ohmic and dielectric losses. In reality, absorption in the dielectric is much larger than the Ohmic loss in the conductor. Therefore, magnetic resonance, which produces antiparallel currents, has been exploited to generate the dielectric loss significantly [16, 18, 23].

MM absorber has been firstly proposed by Landy et al., which had advantage of small size and thin thickness compared with the conventional absorbers [24, 25]. Since Landy et al.'s work, a great number of optimized MM absorber have been proposed for different application areas, such as thermal images [26, 27], solar cell [28], sensor [17] and so on, and the optimization includes multiband [29–34], broad band [20, 35], polarization-insensitive [33, 34, 36, 37] and controllable band [38–40]. Obviously, the MM single-band high absorption is inapplicable in some areas [33]. Therefore, the research on broadband or multi-band high-performance MM absorber is necessary. Nevertheless, it is not easy to combine multi-band MMPAs with high efficiency, since the sensitive perfect absorption conditions are easy to be broken [33]. They recently demonstrated multi-band [33] and broadband [41] by using several kinds of resonators, and even dual-bands by using only one kind of resonator [34]. Furthermore, Ding et al. already reported broadband MM absorbers with quadrangular frustum pyramids using multilayer process and milling method, which showed wide-band absorption in 8–14 GHz [41]. However, the achievement is still a significant issue in the MMPA researches. In spite of numerous studies, many issues remain to be explored, for example, to relax the working conditions and to increase the number of absorption peaks and the absorption bandwidth [42–54], as well as to switch the absorption properties [38, 40] from microwave to infrared (IR) frequencies. Polarization-independent

MM absorber with wide-band high absorption at both low and high frequencies simultaneously has never been reported [1].

Recently, the absorption using MMs has attracted attention due to their tunable and controllable effects originating from alternation of the coupling at the resonance frequency of EM wave [39]. By manipulating the polarization in the MM, the embedded diode has led to switchable absorption in the GHz region [40]. Control of the distance between coupling components allows us to obtain the tunable dual-band perfect absorbers based on extraordinary optical transmission and Fabry-Pérot cavity resonance [38]. The reshaping of MM elements using micro-machined actuators to devise the switchable dual-band absorption at THz frequency has been successfully realized in other work [55]. Furthermore, by utilizing the advantages, controllable absorbers are being developed to promise wide applications in tunable filters, detectors [17, 18], and optical switches [56] in near future.

The interaction between incident EM wave and MMs can be due to multiple reactions [57] and near-field coupling [56] between patterned and continuous metal layers [25]. The usual lattice constant of aforementioned general MM absorbers is 1/3–1/5 of the wavelength of indent EM wave.

EM waves are in various polarization states, and to enhance the absorption the MM absorber should be designed to absorb EM waves independently of the polarization. The endeavor to realize this has been performed. They realized the polarization-independent dual-wide-band MMPAs by using cone-type multilayered structure. It should be noted that, in order to achieve the dual-wide-band absorber, they apply the concept of third magnetic resonance [34]. By comparing between simulation and measurement, the lower frequency band is in excellent coincidence, while the higher frequency one is in slight discrepancy. The low-frequency absorption band turned out to be induced by the fundamental magnetic resonance and the high-frequency one due to the third magnetic resonance, and the polarization independence is presented. They also suggested that the dual broadband is demonstrated even in the infrared and the visible ranges [58]. People can obtain the MM absorption more or less independent of polarization at this moment.

The MM absorbers to be more practical should have the capability to cover large angle of incidence of the EM wave. The endeavor to realize this has been performed. The essential properties, such as low losses [59] or symmetry [60], are pursued to be elucidated and applied. Recently, by taking the advantage of diversification of MM designs, the unit cell, whose shape is based on natural structures, comes to emerge in this field of research [61]. They investigated the dependence of absorption on a wide range of incident angle for both transverse-electric (TE) and transverse-magnetic (TM) polarization [62]. The measured absorption decreased from 99.7 to 88 % when θ increased from 10 to 40°. In addition, the absorption frequency was nearly unchanged according to θ. The slight reduction of absorption according to incident angle can be explained by the coupling between external magnetic field of the incident EM wave and MMPA. Similarly to TE polarization, the single-peak absorption of 98.7 % was obtained at 400 MHz for the normal incidence in case of TM polarization. We keep working to obtain the MM absorption for even larger angle of incidence.

THz or high-frequency MMs have received much attention, since conventional and natural materials hardly response to THz EM waves. THz MMs are beyond this limitation and can interact with THz waves by tailoring unit cells to typical periodicities of tens or hundreds of micrometers [63].

Thus far, though most THz MMs were fabricated on rigid substrates [64–68] (e.g., silicon, sapphire), there have been several studies on flexible THz MMs, which were generally prepared on flexible substrates [25, 69–74], such as polydimethylsiloxane (PDMS), polypropylene (PP), polyethylene terephthalate (PET), and polyethylene naphthalate (PEN). Apart from the advantages of flexible substrates (e.g., high transparency, lightweight, low cost, and portability), they provide more degrees of freedom to manipulate THz responses. Li et al. demonstrated a continuous tunability of resonant frequencies through stretching the PDMS-based MMs [70]. Han et al. fabricated the multilayered PEN MMs and increased the bandwidths of resonant responses by 4.2 times compared with those of single-layered samples [72]. It was noted that electric resonance of SRRs played predominant roles in most flexible microwave, THz, infrared, and visible MMs [75–78]. However, electric-magnetic coupling has not been intensively studied in flexible MMs, which was mostly investigated in rigid MMs.

The effects of bending strain on electric-magnetic coupling were investigated for flexible THz MMs, which were either asymmetric or symmetric. A well-defined plasmon-induced transparency was demonstrated in the asymmetric flexible THz MMs, which were insensitive to the bending strain. However, for the symmetric flexible THz MMs, the bending strain strongly influenced the coupling, giving rise to continuous variations in transmission at a frequency of 1.1 THz [63].

To achieve the perfect absorption, not only the aforementioned ways but the method utilizing electromagnetically-induced transparency (EIT) has been investigated recently. The EIT is a quantum optical phenomenon to make an absorptive medium transparent to a resonant probe field owing to a destructive quantum interference between two pathways induced by a coupling field [79]. Unlike the quantum interference in atomic systems, the coupled components based on different mechanisms can be applied to realize the analogy of EIT in linear classical systems [80]. Most of this plasmonic EIT-like effects at optical frequencies were realized by using the near-field couplings between bright and dark modes [81–85] or between bright and bright modes [86, 87]. A scheme for realizing plasmonic EIT-like effects at optical frequencies in a stacked MM, which consists of two silver rings, was proposed [88]. The two silver rings are excited strongly by an incident wave. Based on Fabry-Pérot resonance coupling between the two resonators, polarization-independent EIT-like effects appear at optical frequencies. Using a planar MM, which consists of two silver strips, they theoretically demonstrated the plasmonic EIT-like spectral response at optical frequencies [84]. The two silver strips serve as the bright modes, and are excited strongly by the incident wave. Based on the weak hybridization between the two bright modes, a highly-dispersive plasmonic EIT-like spectral response appeared in their scheme. Moreover, the group index is higher than that of another scheme which utilizes the strong coupling between the bright and dark modes. The principle for the MM absorption based on EIT will be discussed, and the advances are also mentioned. One of the superiorities of MMs is that the EIT effect can be mimicked in a much easier way by using MMs [89, 90]. A different approach to create a multi-band MMPA was proposed by exploiting the EIT effect [91]. Generally, there is single absorption peak when external EM field excites only one plasmonic resonance. The key idea was that dual-band absorption can be achieved by employing the near-field coupling between bright and dark plasmonic modes even though only one resonance can be directly excited by the EM field. An extended model inducing multi-band absorption is also provided by considering the interaction between dark meta-molecules. Recent researches on MM absorbers and radiators include design of MM-based lenses and antennas, fabrication and measurements of MM structures for antenna applications, design and measurements of MM absorbing materials and screens, industrial applications of MM absorbers and radiators, etc.

The organization of the book is as follows. The Chap. 2 serves as Theoretical Backgrounds to MMPAs and their components. In the Chap. 3, MMPAs operating in different frequency ranges, including MHz, GHz, infrared and optical ranges, are examined. In the Chap. 4, MMPAs, based on EIT has been elucidated. Broadband and tunable MMPAs, and polarization-independent and wide-incident-angle MMPAs are also described in Chaps. 5 and 6, respectively, both of which are important for the practical applications. Perspectives and future works are discussed in the Chap. 7 to end this book.

References

1. J.Y. Rhee, Y.J. Yoo, K.W. Kim, Y.J. Kim, Y.P. Lee, Metamaterial-based perfect absorbers. J. Electromagn. Waves Appl. **28**, 1541 (2014)
2. R.M. Walser, Electromagnetic metamaterials. in *Proceedings of the SPIE*, vol. 4467 (2001), p. 1
3. Defense Advanced Research Projects Agency. http://www.darpa.mil/DARPATech2002/presentations/dso_pdf/speeches/BROWNING.pdf. Cited on 11 Nov 2008
4. N.I. Zheludev, The road ahead for metamaterials. Science **328**, 582 (2010)
5. B. Ung, Metamaterials: A metareview. Available online: http://www.polymtl.ca/phys/doc/art_2_2.pdf. Accessed on 26 Mar 2014
6. D.R. Smith, J.B. Pendry, M.C.K. Wiltshire, Metamaterials and negative refractive index. Science **305**, 788 (2004)
7. N. Seddon, T. Bearpark, Observation of the inverse Doppler effect. Science **302**, 1537 (2003)
8. J.B. Pendry, Negative refraction makes a perfect lens. Phys. Rev. Lett. **85**, 3966 (2000)
9. D. Schurig, J.J. Mock, B.J. Justice, S.A. Cummer, J.B. Pendry, A.F. Starr, D.R. Smith, Metamaterial electromagnetic cloak at microwave frequencies. Science **314**, 977 (2006)
10. V.M. Shalaev, Optical negative-index metamaterials. Nature Photon. **1**, 41 (2007)
11. T. Driscoll, H.T. Kim, B.G. Chae, B.J. Kim, Y.W. Lee, N.M. Jokerst, S. Palit, D.R. Smith, M. Di Ventra, D.N. Basov, Memory metamaterials. Science **325**, 1518 (2009)
12. N. Liu, H. Guo, L. Fu, S. Kaiser, H. Schweizer, H. Giessen, Three-dimensional photonic metamaterials at optical frequencies. Nature Mater. **7**, 31 (2008)
13. P.V. Tuong, Y.J. Yoo, J.W. Park, Y.J. Kim, K.W. Kim, Y.H. Kim, H. Cheong, L.Y. Chen, Y. P. Lee, Multi-plasmon-induced perfect absorption at the third resonance in metamaterials. J. Opt. **17**(12), 125101 (2015)

14. X. Wang, C. Luo, G. Hong, X. Zhao, Metamaterial optical refractive index sensor detected by the naked eye. Appl. Phys. Lett. **102**, 091902 (2013)
15. Y.J. Kim, J.M. Kim, Y.J. Yoo, P.V. Tuong, H. Zheng, J.Y. Rhee, Y. Lee, J, Dual-absorption metamaterial controlled by electromagnetic polarization. Opt. Soc. Am. B **31**, 2744 (2014)
16. J. Hendrickson, J. Guo, B. Zhang, W. Buchwald, R. Soref, Wideband perfect light absorber at midwave infrared using multiplexed metal structures. Opt. Lett. **37**, 371 (2012)
17. N. Liu, M. Mesch, T. Weiss, M. Hentschel, H. Giessen, Infrared perfect absorber and its application as plasmonic sensor. Nano Lett. **10**, 2342 (2010)
18. F.B.P. Niesler, J.K. Gansel, S. Fischbach, M. Wegener, Metamaterial metal-based bolometers. Appl. Phys. Lett. **100**, 203508 (2012)
19. B. Wang, K.H. Teo, T. Nishino, W. Yerazunis, J. Barnwell, J. Zhang, Experiments on wireless power transfer with metamaterials. Appl. Phys. Lett. **98**, 254101 (2011)
20. K. Aydin, V.E. Ferry, R.M. Briggs, H.A. Atwater, Broadband polarization-independent resonant light absorption using ultrathin plasmonic super absorbers. Nature Commun. **2**, 517 (2011)
21. M. Hedayati, F. Faupel, M. Elbahri, Tunable broadband plasmonic perfect absorber at visible frequency. Appl. Phys. A **109**, 769 (2012)
22. A.S. Hall, M. Faryad, G.D. Barber, L. Liu, S. Erten, T.S. Mayer, A. Lakhtakia, T.E. Mallouk, Broadband light absorption with multiple surface plasmon polariton waves excited at the interface of a metallic grating and photonic crystal. ACS Nano **7**, 4995 (2013)
23. P. Tuong, J. Park, V. Lam, W. Jang, S. Nikitov, Y. Lee, Dielectric and Ohmic losses in perfectly-absorbing metamaterials. Opt. Commun. **295**, 17 (2013)
24. N.I. Landy, S. Sajuyigbe, J.J. Mock, D.R. Smith, W.J. Padilla, Perfect metamaterial absorber. Phys. Rev. Lett. **100**, 207402 (2008)
25. Y.J. Yoo, H.Y. Zheng, Y.J. Kim, J.Y. Rhee, J.H. Kang, K.W. Kim, H. Cheong, Y.H. Kim, Y. P. Lee, Flexible and elastic metamaterial absorber for low frequency, based on small-size unit cell. Appl. Phys. Lett. **105**, 041902 (2014)
26. X. Liu, T. Starr, A.F. Starr, W.J. Padilla, Infrared spatial and frequency selective metamaterial with near-unity absorbance. Phys. Rev. Lett. **104**, 207403 (2010)
27. X. Liu, T. Tyler, T. Starr, A.F. Starr, N.M. Jokerst, W.J. Padilla, Taming the blackbody with infrared metamaterials as selective thermal emitters. Phys. Rev. Lett. **107**, 045901 (2011)
28. J. Hao, L. Zhou, M. Qiu, Nearly total absorption of light and heat generation by plasmonic metamaterials. Phys. Rev. B **83**, 165107 (2011)
29. X.Y. Peng, B. Wang, S. Lai, D.H. Zhang, J.H. Teng, Ultrathin multi-band planar metamaterial absorber based on standing wave resonances. Opt. Express. **20**, 27756 (2012)
30. Y. Cheng, Y. Nie, R. Gong, H. Yang, Multi-band metamaterial absorber using cave-cross resonator. Eur. Phys. J. Appl. Phys. **56**, 31301 (2011)
31. Y. Ma, Q. Chen, J. Grant, S.C. Saha, A. Khalid, D.R.S. Cumming, A terahertz polarization insensitive dual band metamaterial absorber. Opt. Lett. **36**, 945 (2011)
32. H. Li, L.H. Yuan, B. Zhou, X.P. Shen, Q. Cheng, T.J. Cui, Ultrathin multiband gigahertz metamaterial absorbers. J. Appl. Phys. **110**, 014909 (2011)
33. J.W. Park, P. Van Tuong, J.Y. Rhee, K.W. Kim, W.H. Jang, E.H. Choi, L.Y. Chen, Y. Lee, Multi-band metamaterial absorber based on the arrangement of donut-type resonators. Opt. Express. **21**, 9691 (2013)
34. Y.J. Yoo, Y.J. Kim, P. Van Tuong, J.Y. Rhee, K.W. Kim, W.H. Jang, Y.H. Kim, H. Cheong, Y.P. Lee, Polarization-independent dual-band perfect absorber utilizing multiple magnetic resonances. Opt. Express. **21**, 32484 (2013)
35. Y. Cui, J. Xu, K. Hung Fung, Y. Jin, A. Kumar, S. He, N.X. Fang, A thin film broadband absorber based on multi-sized nanoantennas. Appl. Phys. Lett. **99**, 253101 (2011)
36. B. Zhang, Y. Zhao, Q. Hao, B. Kiraly, I.C. Khoo, S. Chen, T.J. Huang, Polarization-independent dual-band infrared perfect absorber based on a metal-dielectric-metal elliptical nanodisk array. Opt. Express. **19**, 15221 (2011)

37. N.I. Landy, C.M. Bingham, T. Tyler, N. Jokerst, D.R. Smith, W.J. Padilla, Design, theory, and measurement of a polarization-insensitive absorber for terahertz imaging. Phys. Rev. B **79**, 125104 (2009)
38. H.Y. Zheng, X.R. Jin, J.W. Park, Y.H. Lu, J.Y. Rhee, W.H. Jang, H. Cheong, Y.P. Lee, Tunable dual-band perfect absorbers based on extraordinary optical transmission and Fabry-Perot cavity resonance. Opt. Express. **20**, 24002 (2012)
39. P.V. Tuong, J.W. Park, J.Y. Rhee, K.W. Kim, W.H. Jang, H. Cheong, Y.P. Lee, Polarization-insensitive and polarization-controlled dual-band absorption in metamaterials. Appl. Phys. Lett. **102**, 081122 (2013)
40. B. Zhu, Y. Feng, J. Zhao, C. Huang, T. Jiang, Switchable metamaterial reflector/absorber for different polarized electromagnetic waves. Appl. Phys. Lett. **97**, 051906 (2010)
41. F. Ding, Y. Cui, X. Ge, Y. Jin, S. He, Ultra-broadband microwave metamaterial absorber. Appl. Phys. Lett. **100**, 103506 (2012)
42. P. Tuong, J. Park, V. Lam, K. Kim, H. Cheong, W. Jang, Y. Lee, Simplified perfect absorber structure. Comput. Mater. Sci. **61**, 243 (2012)
43. S. Chen, H. Cheng, H. Yang, J. Li, X. Duan, C. Gu, J. Tian, Polarization insensitive and omnidirectional broadband near perfect planar metamaterial absorber in the near infrared regime. Appl. Phys. Lett. **99**, 253104 (2011)
44. X. Shen, T.J. Cui, J. Zhao, H.F. Ma, W.X. Jiang, H. Li, Polarization-independent wide-angle triple-band metamaterial absorber. Opt. Express. **19**, 9401 (2011)
45. J. Sun, L. Liu, G. Dong, J. Zhou, An extremely broad band metamaterial absorber based on destructive interference. Opt. Express. **19**, 21155 (2011)
46. H.X. Xu, G.M. Wang, M.Q. Qi, J.G. Liang, J.Q. Gong, Z.M. Xu, Triple-band polarization-insensitive wide-angle ultra-miniature metamaterial transmission line absorber. Phys. Rev. B **86**, 205104 (2012)
47. H. Cheng, S. Chen, H. Yang, J. Li, X. An, C. Gu, J. Tian, A polarization insensitive and wide-angle dual-band nearly perfect absorber in the infrared regime. J. Opt. **14**, 085102 (2012)
48. J. Zhong, Y. Huang, G. Wen, H. Sun, P. Wang, O. Gordon, Single-/dual-band metamaterial absorber based on cross-circular-loop resonator with shorted stubs. Appl. Phys. A **108**, 329 (2012)
49. H.M. Lee, J.C. Wu, A wide-angle dual-band infrared perfect absorber based on metal–dielectric–metal split square-ring and square array. J. Phys. D Appl. Phys. **45**, 205101 (2012)
50. H. Tao, N.I. Landy, C.M. Bingham, X. Zhang, R.D. Averitt, W.J. Padilla, A metamaterial absorber for the terahertz regime: Design, fabrication and characterization. Opt. Express. **16**, 7181 (2008)
51. L. Huang, H. Chen, Multi band and polarization insensitive metamaterial absorber. Prog. Electromagnet. Res. **113**, 103 (2011)
52. L. Zhu, F.Y. Meng, J.H. Fu, Q. Wu, J. Hua, Multi-band slow light metamaterial. Opt. Express. **20**, 4494 (2012)
53. F.Y. Meng, Q. Wu, D. Erni, K. Wu, J.C. Lee, Polarization-independent metamaterial analog of electromagnetically induced transparency for a refractive-index-based sensor. IEEE. Trans, Microw. Theory Tech. **60**, 3013 (2012)
54. D. Brazhnikov, A. Taichenachev, V. Yudin, Polarization method for controlling a sign of electromagnetically-induced transparency/absorption resonances. Eur. Phys. J. D **63**, 315 (2011)
55. W. Zhang, A.Q. Liu, W.M. Zhu, E.P. Li, H. Tanoto, Q.Y. Wu, J.H. Teng, X.H. Zhang, M.L. J. Tsai, G.Q. Lo, D.L. Kwong, Micromachined switchable metamaterial with dual resonance. Appl. Phys. Lett. **101**, 151902 (2012)
56. H.T. Chen, W.J. Padilla, J.M.O. Zide, S.R. Bank, A.C. Gossard, A.J. Taylor, R.D. Averitt, Ultrafast optical switching of terahertz metamaterials fabricated on ErAs/GaAs nanoisland superlattices. Opt. Lett. **32**, 1620 (2007)
57. L. Huang, D.R. Chowdhury, S. Ramani, M.T. Reiten, S.N. Luo, A.K. Azad, A.J. Taylor, H.T. Chen, Impact of resonator geometry and its coupling with ground plane on ultrathin metamaterial perfect absorbers. Appl. Phys. Lett. **101**, 101102 (2012)

58. Y.J. Kim, Y.J. Yoo, K.W. Kim, J.Y. Rhee, Y.H. Kim, Y. Lee, Dual broadband metamaterial absorber. Opt. Express. **23**, 3861 (2015)
59. A. Boltasseva, H.A. Atwater, Low-loss plasmonic metamaterials. Science **331**, 290 (2011)
60. K.B. Alici, E. Ozbay, A planar metamaterial: Polarization independent fishnet structure. Photon. Nanostruct. **6**, 102 (2008)
61. W. Zhu, X. Zhao, B. Gong, L. Liu, B. Su, Optical metamaterial absorber based on leaf-shaped cells. Appl. Phys. A **102**, 147 (2011)
62. B.X. Khuyen, B.S. Tung, N.V. Dung, Y.J. Yoo, Y.J. Kim, K.W. Kim, V.D. Lam, J.G. Yang, Y. Lee, J. Size-efficient metamaterial absorber at low frequencies: Design, fabrication, and characterization. Appl. Phys. **117**, 243105 (2015)
63. L. Zheng, X. Sun, H. Xu, Y. Lu, Y. Lee, J. Rhee, W. Song, Strain sensitivity of electric-magnetic coupling in flexible terahertz metamaterials. Plasmonics. **10**, 1331 (2015)
64. H.T. Chen, W.J. Padilla, J.M.O. Zide, A.C. Gossard, A.J. Taylor, R.D. Averitt, Active terahertz metamaterial devices. Nature **444**, 597 (2006)
65. Y. Urzhumov, J.S. Lee, T. Tyler, S. Dhar, V. Nguyen, N.M. Jokerst, P. Schmalenberg, D.R. Smith, Electronically reconfigurable metal-on-silicon metamaterial. Phys. Rev. B **86**, 075112 (2012)
66. F. Valmorra, G. Scalari, C. Maissen, W. Fu, C. Schönenberger, J.W. Choi, H.G. Park, M. Beck, J. Faist, Low-bias active control of terahertz waves by coupling large-area CVD graphene to a terahertz metamaterial. Nano Lett. **13**, 3193 (2013)
67. K. Fan, A.C. Strikwerda, X. Zhang, R.D. Averitt, Three-dimensional broadband tunable terahertz metamaterials. Phys. Rev. B **87**, 161104 (2013)
68. J. Gu, R. Singh, X. Liu, X. Zhang, Y. Ma, S. Zhang, S.A. Maier, Z. Tian, A.K. Azad, H.T. Chen, A.J. Taylor, J. Han, W. Zhang, Active control of electromagnetically induced transparency analogue in terahertz metamaterials. Nature Commun. **3**, 1151 (2012)
69. J. Li, C.M. Shah, W. Withayachumnankul, B.S.Y. Ung, A. Mitchell, S. Sriram, M. Bhaskaran, S. Chang, D. Abbott, Flexible terahertz metamaterials for dual-axis strain sensing. Opt. Lett. **38**, 2104 (2013)
70. J. Li, C.M. Shah, W. Withayachumnankul, B.S.Y. Ung, A. Mitchell, S. Sriram, M. Bhaskaran, S. Chang, D. Abbott, Mechanically tunable terahertz metamaterials. Appl. Phys. Lett. **102**, 121101 (2013)
71. X. Wen, G. Li, J. Zhang, Q. Zhang, B. Peng, L.M. Wong, S. Wang, Q. Xiong, Transparent free-standing metamaterials and their applications in surface-enhanced Raman scattering. Nanoscale **6**, 132 (2014)
72. R. Ortuno, C. Garca-Meca, A. Martnez, Terahertz metamaterials on flexible polypropylene substrate. Plasmonics **9**, 1143 (2014)
73. N.R. Han, Z.C. Chen, C.S. Lim, B. Ng, M.H. Hong, Broadband multi-layer terahertz metamaterials fabrication and characterization on flexible substrates. Opt. Express. **19**, 6990 (2011)
74. J.M. Woo, D. Kim, S. Hussain, J.H. Jang, Low-loss flexible bilayer metamaterials in THz regime. Opt. Express. **22**, 2289 (2014)
75. P.K. Singh, K.A. Korolev, M.N. Afsar, S. Sonkusale, Sonkusale, Single and dual band 77/95/110 GHz metamaterial a sor ers on flexi le polyi ide su strate. Appl. Phys. Lett. **99**, 264101 (2011)
76. D.H. Kim, D.S. Kim, S. Hwang, J.H. Jang, Surface relief structures for a flexible broadband terahertz absorber. Opt. Express. **20**, 16815 (2012)
77. Z.H. Jiang, S. Yun, F. Toor, D.H. Werner, T.S. Mayer, Conformal dual-band near-perfectly absorbing mid-infrared metamaterial coating. ACS Nano **5**, 4641 (2011)
78. A. Di Falco, Y. Zhao, A. Al, Optical metasurfaces with robust angular response on flexible substrates. Appl. Phys. Lett. **99**, 163110 (2011)
79. M. Fleischhauer, A. Imamoglu, J.P. Marangos, Electromagnetically induced transparency: Optics in Coherent Media. Rev. Mod. Phys. **77**, 633 (2005)

80. Y. Lu, H. Xu, J.Y. Rhee, W.H. Jang, B.S. Ham, Y. Lee, Magnetic plasmon resonance: Underlying route to plasmonic electromagnetically induced transparency in metamaterials. Phys. Rev. B **82**, 195112 (2010)
81. N. Liu, S. Kaiser, H. Giessen, Magnetoinductive and electroinductive coupling in plasmonic metamaterial molecules. Adv. Mater. **20**, 4521 (2008)
82. N. Liu, L. Langguth, T. Weiss, J. Kästel, M. Fleischhauer, T. Pfau, H. Giessen, Plasmonic analogue of electromagnetically induced transparency at the Drude damping limit. Nature Mater. **8**, 758 (2009)
83. X. Jin, Y. Lu, H. Zheng, Y. Lee, J.Y. Rhee, W.H. Jang, Plasmonic electromagnetically-induced transparency in symmetric structures. Opt. Express. **18**, 13396 (2010)
84. Y. Lu, J.Y. Rhee, W.H. Jang, Y.P. Lee, Active manipulation of plasmonic electromagnetically-induced transparency based on magnetic plasmon resonance. Opt. Express. **18**, 20912 (2010)
85. N. Liu, T. Weiss, M. Mesch, L. Langguth, U. Eigenthaler, M. Hirscher, C. Snnichsen, H. Giessen, Planar metamaterial analogue of electromagnetically induced transparency for plasmonic sensing. Nano Lett. **10**, 1103 (2010)
86. X.R. Jin, J. Park, H. Zheng, S. Lee, Y. Lee, J.Y. Rhee, K.W. Kim, H.S. Cheong, W.H. Jang, Highly-dispersive transparency at optical frequencies in planar metamaterials based on two-bright-mode coupling. Opt. Express. **19**, 21652 (2011)
87. R.D. Kekatpure, E.S. Barnard, W. Cai, M.L. Brongersma, Phase-coupled plasmon-induced transparency. Phys. Rev. Lett. **104**, 243902 (2010)
88. X.R. Jin, Y.Q. Zhang, S. Zhang, Y. Lee, J.Y. Rhee, Polarization-independent electromagnetically induced transparency-like effects in stacked metamaterials based on Fabry–Perot resonance. J. Opt. **15**, 125104 (2013)
89. S. Zhang, D.A. Genov, Y. Wang, M. Liu, X. Zhang, Plasmon-induced transparency in metamaterials. Phys. Rev. Lett. **101**, 047401 (2008)
90. V.T.T. Thuy, N.T. Tung, J.W. Park, V.D. Lam, Y.P. Lee, J.Y. Rhee, Highly dispersive transparency in coupled Metamaterials. J. Opt. **12**, 115102 (2010)
91. B.S. Tung, B.X. Khuyen, N.V. Dung, V.D. Lam, Y.H. Kim, H. Cheong,, Y.P. Lee, Multi-band near-perfect absorption via the resonance excitation of dark meta-molecules. Opt. Commun. (2015)

Chapter 2
Theoretical Backgrounds

Abstract In this chapter, some theoretical aspects of propagation of electromagnetic (EM) waves in matters and their interaction with matters, which are essential to understand the phenomena occurring in metamaterial perfect absorber (MMPA) and to design and/or optimize the MMPA structures, will be presented. A brief introduction of broadband and resonant absorbers is provided. After a brief discussion on MMPA is given, the effective-medium approximation, which is essential for extracting various parameters from the simulated or the measured spectrum of MMPA, and its validity limit will be discussed. The equivalent-circuit theory and the transmission-line theory, will be briefly presented, and the introduction of several numerical techniques, such as finite-difference time-domain method, finite-element method and transfer-matrix method, will conclude this chapter.

2.1 Interaction of Electromagnetic Waves with Matter

It is very well-known that the equations for describing the propagation of electromagnetic (EM) waves through media and their interaction with matter are the Maxwell's equations, and they are the main topics of many textbooks, such as Jackson [1], Brau [2], Reitz et al. [3], and so on. When an EM wave travels through a medium, the propagation of the EM wave is governed by the Maxwell's equations written as[1]

$$\begin{array}{ll} \nabla \cdot \mathbf{D} = \rho & \text{(Gauss law)}, \\ \nabla \cdot \mathbf{B} = 0 & \text{(Absence of magnetic monopole)}, \\ \nabla \times \mathbf{E} = -\frac{\partial \mathbf{B}}{\partial t} & \text{(Faraday's law)}, \\ \nabla \times \mathbf{H} = \mathbf{J} + \frac{\partial \mathbf{D}}{\partial t} & \text{(Maxwell-Ampere's law)}, \end{array} \tag{2.1}$$

[1]SI units will be used throughout this chapter.

Y.P. Lee et al., *Metamaterials for Perfect Absorption*,
Springer Series in Materials Science 236, DOI 10.1007/978-981-10-0105-5_2

where **E** and **D** are electric field and electric displacement, respectively, **H** and **B** are magnetic field and magnetic induction, respectively. ρ and **J** are charge and current densities, respectively. **D**, **B**, and **J** satisfy the constitutive (or material) relations:

$$\begin{aligned} \mathbf{D} &= \varepsilon \mathbf{E}, \\ \mathbf{B} &= \mu \mathbf{H}, \\ \mathbf{J} &= \sigma \mathbf{E}, \end{aligned} \tag{2.2}$$

where ε is electric permittivity, μ is magnetic permeability, and σ is the electrical conductivity. If the medium is anisotropic, these three quantities (ε, μ and σ) come to be second-rank tensors; however, we are dealing with the isotropic media only, otherwise specifically stated. These quantities are also dependent upon their respective fields, i.e.,

$$\begin{aligned} \varepsilon &= \varepsilon(\mathbf{E}), \\ \mu &= \mu(\mathbf{H}), \\ \sigma &= \sigma(\mathbf{E}). \end{aligned} \tag{3.3}$$

If we expand, for instance, the permittivity in terms of electric field as $\varepsilon = \varepsilon(\mathbf{E}) = \varepsilon^0 + \varepsilon^1 E + \varepsilon^2 E^2 + \cdots$,[2] the second and higher terms are related to the nonlinear effects of EM wave. In most media the first term ε^0 dominates the permittivity unless the electric field is too strong to neglect the higher-order terms. In these cases, consequently, $\varepsilon = \varepsilon^0$. If a medium is under the condition that the nonlinear effects are negligible, we call that medium a linear medium. Again, in this text, we are dealing with linear media only.

It is not difficult to derive the wave equation for EM wave thorough the Maxwell's equations. Here, the most simple case will be presented. Consider an EM wave with a definite (angular) frequency, ω, and wave vector, **k**, propagating through vacuum (or free space). In vacuum $\rho = 0$ and $\mathbf{J} = 0$. Furthermore, $\mathbf{D} = \varepsilon_0 \mathbf{E}$ and $\mathbf{B} = \mu_0 \mathbf{H}$, where ε_0 and μ_0 are the electrical permittivity and the magnetic permeability of vacuum, respectively. In vacuum, therefore, the Maxwell's equations (2.1) become

$$\begin{aligned} \nabla \cdot \mathbf{E} &= 0, \\ \nabla \cdot \mathbf{B} &= 0, \\ \nabla \times \mathbf{E} &= -\frac{\partial \mathbf{B}}{\partial t}, \\ \nabla \times \mathbf{B} &= \varepsilon_0 \mu_0 \frac{\partial \mathbf{E}}{\partial t}. \end{aligned} \tag{2.4}$$

[2]It should be noted here that the superscripts of ε's are the indices, not the exponents.

The first two conditions require $\mathbf{E} \perp \mathbf{k}$ and $\mathbf{B} \perp \mathbf{k}$, which is called the transversality of EM wave in vacuum. By taking the curl of both sides of third formula of (2.4) we can get

$$\nabla \times \nabla \times \mathbf{E} = \nabla \times \left(-\frac{\partial \mathbf{B}}{\partial t}\right) = -\frac{\partial}{\partial t}(\nabla \times \mathbf{B}) = -\varepsilon_0 \mu_0 \frac{\partial^2 \mathbf{E}}{\partial t^2}. \tag{2.5}$$

Since

$$\nabla \times \nabla \times \mathbf{E} = \nabla(\nabla \cdot \mathbf{E}) - \nabla^2 \mathbf{E}, \tag{2.6}$$

the wave equation for the electric field

$$\nabla^2 \mathbf{E} = \varepsilon_0 \mu_0 \frac{\partial^2 \mathbf{E}}{\partial t^2} \tag{2.7}$$

is easily derivable by using the Gauss law [the first formula of (2.4)]. A similar procedure can be applied for obtaining the wave equation for the magnetic field as

$$\nabla^2 \mathbf{B} = \varepsilon_0 \mu_0 \frac{\partial^2 \mathbf{B}}{\partial t^2}. \tag{2.8}$$

The coefficients of the right-hand side of both wave equations correspond to the reciprocal of square of the wave velocity as

$$\varepsilon_0 \mu_0 = \frac{1}{c^2}, \tag{2.9}$$

where c is the speed of light in vacuum.

If light travels inside a matter, the wave equation is altered because $\rho \neq 0$ and $\mathbf{J} \neq 0$. The wave equation for the electric field becomes

$$\frac{1}{c^2}\frac{\partial^2 \mathbf{E}}{\partial t^2} - \nabla^2 \mathbf{E} = -\left(\frac{1}{\varepsilon_0}\nabla \rho + \mu_0 \frac{\partial \mathbf{J}}{\partial t}\right) \tag{2.10}$$

and that for the magnetic field becomes

$$\frac{1}{c^2}\frac{\partial^2 \mathbf{B}}{\partial t^2} - \nabla^2 \mathbf{B} = \mu_0 \nabla \times \mathbf{J}. \tag{2.11}$$

By introducing the scalar (φ) and vector ($\mathbf{A}$) potentials, such as

$$\mathbf{E} = -\nabla \varphi - \frac{\partial \mathbf{A}}{\partial t} \quad \text{and} \quad \mathbf{B} = \nabla \times \mathbf{A}, \tag{2.12}$$

the Gauss law for the electricity becomes

$$\nabla^2\varphi + \frac{\partial}{\partial t}(\nabla \cdot \mathbf{A}) = -\frac{\rho}{\varepsilon_0} \tag{2.13}$$

and the Ampère-Maxwell law becomes

$$\nabla^2\mathbf{A} - \frac{1}{c^2}\frac{\partial^2\mathbf{A}}{\partial t^2} - \nabla\left(\frac{1}{c^2}\frac{\partial\varphi}{\partial t} + \nabla \cdot \mathbf{A}\right) = -\mu_0\mathbf{J}. \tag{2.14}$$

Since the potentials are not unique, but have gauge freedom, by using the Lorentz gauge

$$\frac{1}{c^2}\frac{\partial\varphi}{\partial t} + \nabla \cdot \mathbf{A} = 0, \tag{2.15}$$

two following wave equations can be derived:

$$\nabla^2\varphi - \frac{1}{c^2}\frac{\partial^2\varphi}{\partial t^2} = -\frac{\rho}{\varepsilon_0} \tag{2.16}$$

and

$$\nabla^2\mathbf{A} - \frac{1}{c^2}\frac{\partial^2\mathbf{A}}{\partial t^2} = -\mu_0\mathbf{J}. \tag{2.17}$$

Solutions for these wave equations are

$$\varphi(\mathbf{r}, t) = \frac{1}{4\pi\varepsilon_0}\int \frac{\rho(\mathbf{r}', t')}{|\mathbf{r} - \mathbf{r}'|}\delta(\tilde{t})d^3r'dt' \tag{2.18}$$

and

$$\mathbf{A}(\mathbf{r}, t) = \frac{\mu_0}{4\pi}\int \frac{\mathbf{J}(\mathbf{r}', t')}{|\mathbf{r} - \mathbf{r}'|}\delta(\tilde{t})d^3r'dt', \tag{2.19}$$

where $\delta(\tilde{t})$ is the Dirac δ-function with $\tilde{t} \equiv t' + \frac{|\mathbf{r}-\mathbf{r}'|}{nc} - t$. n is the refractive index of the medium. The potentials at a certain point $\mathbf{r}$ and a certain time t are determined by the stimuli originating from all source points $\mathbf{r}'$'s. If a stimulus of a particular source point $\mathbf{r}'$, which determines the potential at a certain point $\mathbf{r}$ and a certain time t, leaves at a particular time t', the stimulus should reach the point $\mathbf{r}$ at a later (or retarded) time $t = t' + \frac{|\mathbf{r}-\mathbf{r}'|}{nc}$. These potentials are, therefore, called 'retarded' potentials. The Dirac δ-function assures this condition.

2.1.1 Boundary Conditions and Fresnel Equations

In real world there is no infinite medium. It implies that any medium should have own boundaries, which are in contact with other media. Consider a ray of EM wave traveling along a medium (we will refer to this medium as the first medium) which eventually reach its planar boundary. When the EM wave hits the boundary, some part of incident EM wave is reflected and the other portion is transmitted (or refracted) toward the other medium (we will refer to this medium as the second medium). Let the planar boundary be the xy-plane and the z-axis be normal to the interface. Therefore, the first (second) medium is located at $z<0$ ($z>0$). Assume that the first and second media have the permittivity and permeability of ε, μ and ε', μ', respectively. For the time being, assume that all these 4 quantities are real and positive. If the light hits the boundary obliquely, the incident, the reflected and the transmitted rays are coplanar, and we call this plane as the plane of incidence (PoI). Let $\mathbf{E}_i$, $\mathbf{E}_t$ and $\mathbf{E}_r$ be the electric fields of the incident, the transmitted and the reflected rays, respectively. The same rule is applied to the other fields and the wave vector. The angle of incidence (ϕ_i) and that of reflection (ϕ_r) are the angle between $-\hat{\mathbf{z}}$ ($\hat{\mathbf{z}}$ is the primitive vector along the z-axis) and the incident and the reflected rays, respectively, while the angle of refraction (ϕ_t) is the angle between $\hat{\mathbf{z}}$ and the transmitted ray.

The existence of boundary conditions (BCs) requires that the spatio-temporal variation of all fields must be identical at all time and at all points on the boundary ($z=0$). Therefore, the phase factors of all fields should be the same. For the simplicity consider a plane waves whose phase factor is $e^{i(\mathbf{k}\cdot\mathbf{r}-\omega t)}$. The first set of BCs can be obtained by this equivalence of phase factor. In the first medium, the relation

$$\mathbf{k}_i \cdot \mathbf{r}\big|_{z=0} = \mathbf{k}_r \cdot \mathbf{r}\big|_{z=0}, \tag{2.20}$$

where $\mathbf{k}_i$ and $\mathbf{k}_r$ are the wave vectors of incident and reflected waves, respectively, holds. Since $k_i = k_r$, this relation yields the law of reflection, i.e., $\phi_i = \phi_r$. Another relation

$$\mathbf{k}_i \cdot \mathbf{r}\big|_{z=0} = \mathbf{k}_t \cdot \mathbf{r}\big|_{z=0}, \tag{2.21}$$

where $\mathbf{k}_t$ is the wave vectors of transmitted wave, also holds. Since $\sqrt{\varepsilon\mu}k_i = \sqrt{\varepsilon'\mu'}k_t$ or $nk_i = n'k_t$, where n and n' are the refractive indices of the first and the second media, respectively, this relation yields the law of refraction or Snell's law, i.e., $n\sin\phi_i = n'\sin\phi_t$. It is well-known that, when light travels through a medium and hit its boundary, the light changes its direction of propagation in the second medium. It is, however, only a part of the story. The more important point is that the refraction (bending of the propagation direction) is a result of change of the speed of light in different medium. More specifically, the speed of light in a medium is inversely proportional to its refractive index, i.e., (the speed of light in a medium) = (the speed of light in vacuum)/(refractive index).

The second set of BCs can be obtained from the continuity of some components of fields across the boundary. The tangential (to the boundary) components of **E** and **H** should be continuous, and the normal (to the boundary) components of **D** and **B** should be continuous. Therefore,

$$\begin{aligned}
[\mathbf{E}_{0i} + \mathbf{E}_{0r} - \mathbf{E}_{0t}] \times \hat{\mathbf{z}} &= 0\text{: tangential component of } \mathbf{E},\\
[\varepsilon(\mathbf{E}_{0i} + \mathbf{E}_{0r}) + \varepsilon'\mathbf{E}_{0t}] \cdot \hat{\mathbf{z}} &= 0\text{: normal component of } \mathbf{D},\\
\left[\frac{1}{\mu}(\mathbf{k}_i \times \mathbf{E}_{0i} + \mathbf{k}_i \times \mathbf{E}_{0r}) - \frac{1}{\mu'}\mathbf{k}_t \times \mathbf{E}_{0t}\right] \times \hat{\mathbf{z}} &= 0\text{: tangential component of } \mathbf{H},\\
[\mathbf{k}_i \times \mathbf{E}_{0i} + \mathbf{k}_i \times \mathbf{E}_{0r} - \mathbf{k}_t \times \mathbf{E}_{0t}] \cdot \hat{\mathbf{z}} &= 0\text{: normal component of } \mathbf{B}.
\end{aligned} \tag{2.22}$$

If **E** is perpendicular to PoI, the first condition of (2.22) yields,

$$E_{0i} + E_{0r} - E_{0t} = 0 \tag{2.23}$$

and the third condition yields

$$\sqrt{\frac{\varepsilon}{\mu}}(E_{0i} - E_{0r}) \cos\phi_i - \sqrt{\frac{\varepsilon'}{\mu'}}E_{0t} \cos\phi_t = 0. \tag{2.24}$$

From these two relations we can get the reflection coefficient as

$$r_s \equiv \frac{E_{0r}}{E_{0i}} = \frac{2n\cos\phi_i}{n\cos\phi_i + \frac{\mu}{\mu'}\sqrt{n'^2 - n^2\sin^2\phi_i}} \tag{2.25}$$

and the transmission coefficient as

$$t_s \equiv \frac{E_{0t}}{E_{0i}} = \frac{n\cos\phi_i - \frac{\mu}{\mu'}\sqrt{n'^2 - n^2\sin^2\phi_i}}{n\cos\phi_i + \frac{\mu}{\mu'}\sqrt{n'^2 - n^2\sin^2\phi_i}}. \tag{2.26}$$

Here, the subscript '*s*' stands for senkrecht, German for perpendicular. The linearly-polarized wave with the electric field perpendicular to PoI is called *s*-wave or transverse-electric (TE) wave.

If **E** is parallel to PoI, the second condition of (2.22) yields,

$$\sqrt{\frac{\varepsilon}{\mu}}(E_{0i} + E_{0r}) - \sqrt{\frac{\varepsilon'}{\mu'}}E_{0t} = 0 \tag{2.27}$$

and the fourth condition yields

$$(E_{0i} - E_{0r})\cos\phi_i - E_{0t}\cos\phi_t = 0. \tag{2.28}$$

From these two relations we can get the reflection coefficient as

$$r_p \equiv \frac{E_{0r}}{E_{0i}} = \frac{2nn'\cos\phi_i}{\frac{\mu}{\mu'}n'^2\cos\phi_i + n\sqrt{n'^2 - n^2\sin^2\phi_i}} \tag{2.29}$$

and the transmission coefficient as

$$t_p \equiv \frac{E_{0t}}{E_{0i}} = \frac{\frac{\mu}{\mu'}n'^2\cos\phi_i - n\sqrt{n'^2 - n^2\sin^2\phi_i}}{\frac{\mu}{\mu'}n'^2\cos\phi_i + n\sqrt{n'^2 - n^2\sin^2\phi_i}}. \tag{2.30}$$

Here, the subscript '*p*' stands for parallel. The linearly-polarized wave with the electric field parallel to PoI is called *p*-wave or transverse-magnetic (TM) wave. These 4 equations are called Fresnel's equations.

2.1.2 Dispersion Relations

In most general case for an isotropic and linear medium, the refractive index depends on the wavelength (or frequency) of traveling EM wave. If the refractive index of medium is frequency-dependent, we call the medium dispersive and the frequency dependence of refractive index, which is called the dispersion relation such that

$$\omega^2\tilde{\varepsilon}\tilde{\mu} - \tilde{k}^2 = 0, \tag{2.31}$$

where ω and $\mathbf{k}$ are the (angular) frequency and the wave vector of the EM wave respectively, holds. Here, $\tilde{\varepsilon}$, for example, is used instead of ε itself to indicate that the permittivity is, in general, a complex quantity in the dispersion relation. In fact,

$$\tilde{\varepsilon} = \varepsilon + i\frac{\sigma}{\omega} \equiv \varepsilon_1 + i\varepsilon_2, \tag{2.32}$$

where ε_1 and ε_2 are the real and imaginary parts of the (complex) permittivity. In general cases, the permeability can also be complex. If the permittivity and/or permeability are complex, the wave vector should also be complex. In (2.31) we put a tilde ($\sim$) on top of the permittivity, the permeability and the wave vector to explicitly specify that these quantities are complex. We will put a tilde on top of any physical quantity which is complex and to explicitly be differentiated from their corresponding real quantities. It should be particularly noted here that the sign of the imaginary part of any linear response function should depend on the choice of the sign convention of time dependence of EM wave. If the time dependence of EM wave is $e^{-i\omega t}$ ($e^{i\omega t}$) the sign of the imaginary part of any linear response function should be + (−) because any passive medium is absorptive, and any EM wave propagating through an absorptive medium should be attenuated.

The simplest case is the isotropic, non-magnetic, dielectric medium in which both the permittivity and the permeability are real and positive. In this case the refractive index n satisfies the relation

$$\frac{c}{n} = \frac{1}{\sqrt{\varepsilon\mu}}. \tag{2.33}$$

In vacuum, the wave number is linearly proportional to the frequency in whole frequency range. As can be seen in (2.32), however, the permittivity and the permeability themselves are frequency-dependent. This frequency dependence is the results of the non-locality of time. When the electric displacement is connected to the electric field, **D** at time t depends on **E** at all times other than t. The non-locality of time, together with the causality, requires following Kramers-Kronig (KK) relations;

$$\begin{aligned} \varepsilon_1(\omega) &= \varepsilon_0 + \frac{2}{\pi} P \int_0^\infty \frac{\omega' \varepsilon_2(\omega')}{\omega'^2 - \omega^2} d\omega', \\ \varepsilon_2(\omega) &= -\frac{2\omega}{\pi} P \int_0^\infty \frac{\varepsilon_1(\omega') - \varepsilon_0}{\omega'^2 - \omega^2} d\omega', \end{aligned} \tag{2.34}$$

where P means the principal part. The KK relations imply that if any medium is dispersive, it is also absorptive.

Any response function of linear medium should satisfy the KK relations. The complex refractive index can be written as $\tilde{n} = n + i\kappa$, where n is the ordinary refractive index and κ is the extinction coefficient. The complex refractive index also satisfies the following KK relations;

$$\begin{aligned} n(\omega) &= 1 + \frac{2}{\pi} P \int_0^\infty \frac{\omega' \kappa(\omega')}{\omega'^2 - \omega^2} d\omega', \\ \kappa(\omega) &= -\frac{2\omega}{\pi} P \int_0^\infty \frac{n(\omega') - 1}{\omega'^2 - \omega^2} d\omega'. \end{aligned} \tag{2.35}$$

It is tempting to apply the same KK relations for the complex reflection coefficient. By taking the natural log of the complex reflectance as $\ln \tilde{r}(\omega) = \frac{1}{2}R(\omega) + i\theta(\omega)$ because $\tilde{r}(\omega) = \sqrt{R(\omega)}e^{i\theta(\omega)}$, where $R(\omega) = |\tilde{r}(\omega)|^2$ and $\tan\theta(\omega) = \mathrm{Im}[\tilde{r}(\omega)]/\mathrm{Re}[\tilde{r}(\omega)]$. Therefore, the following KK relation

$$\theta(\omega) = -\frac{\omega}{\pi}P\int_{0}^{\infty}\frac{\ln R(\omega)}{\omega'^2 - \omega^2}d\omega' \tag{2.36}$$

can be used. $R(\omega)$ is the reflectance, which is defined as the ratio of power reflected to the power incident, and $\theta(\omega)$ is the phase change upon reflection. Although in experiment the reflectivity, which is the absolute square of reflection coefficient, can be easily measured, the measured reflectance is not possible to compare with the theory. The reflectance is a function of complex permittivity. It implies that the reflectance can be theoretically found if both ε_1 and ε_2 are known. Only ε_2, or equivalently the optical conductivity, can be theoretically obtained through the band-structure calculation of material. Therefore, it is necessary to find the phase change $[\theta(\omega)]$ upon reflection to compare the measured reflectivity with the theoretical one. We cannot apply the relationship, (2.36) because of two reasons. First, the $\ln \tilde{r}(\omega)$ is not a response function of material. Second, we cannot make a complete contour for the complex integration needed to derive the KK relation, because $\ln R(\omega)$ is unbounded asymptotically, i.e., $\lim_{\omega\to\infty} \ln R(\omega) \to -\infty$ or $\lim_{\omega\to\infty} R(\omega) \to 0$. In practice; however, it is not impossible to use the above relationship as a reasonable approximation for normal incidence. In actual measurement of reflectivity, the angle of incidence is not zero, but near zero. Therefore, it is necessary to correct the small angular deviation. Furthermore, it is impossible to measure the whole range of frequency to apply the KK relation. Then it is inevitable to extrapolate the data to outside the measured range. Although these two facts mean that some degree of error is unavoidable, the approximate correction and a careful extrapolation affect only the magnitudes of optical constants, but do not significantly change the spectral shape. Ellipsometry is, therefore, the powerful tool to measure the optical response of material since it can measure both ε_1 and ε_2 simultaneously.

Until 1967, it was naturally believed that the refractive index was positive. Veselago [4] was the very first physicist who realized that the refractive index could be negative if both permittivity and permeability were negative. If the refractive index is negative the phase velocity is opposite to the direction of propagation of energy, which is determined by the Poynting vector ($\mathbf{S} \equiv \mathbf{E} \times \mathbf{H}$), implying that the energy and the phase of EM wave propagate oppositely in the medium with negative values of ε and μ. **E**, **H**, and **k** form a left-hand triad in medium with negative ε and μ, therefore, the medium with negative ε and μ is called a left-handed medium. It is not possible; however, to find natural materials with simultaneously negative ε and μ. Although ε_1 could be negative for metals in the frequency below the plasma frequency ω_p, defined as

$$\omega_p = \sqrt{\frac{ne^2}{m^* \varepsilon_0}}, \tag{2.37}$$

where e is the electronic charge, n the charge density, m^* the effective mass, the permittivity of metal in the frequency range below the plasma frequency does not have a real value, but a complex value. It was not possible to find materials with negative real part of the permeability until Pendry et al. [5] theoretically proposed, and Smith et al. [6] experimentally realized split-ring resonators as the component with negative permeability. One may argue that a diamagnetic medium can have negative μ; however, the diamagnetic behavior occurs under a static condition ($\omega = 0$). In any case, a natural material which possesses simultaneously negative values of ε and μ has not been found or discovered yet.

2.2 Perfect Electromagnetic-Wave Absorbers

Perfect absorbers are needed in many fields of technology, such as various EM-wave detectors, bolometers, solar-energy harvesting, and so on. Ruck et al. [7] classified perfect EM-wave absorbers into two groups; (i) broadband absorbers and (ii) resonant absorbers. So far, the resonant absorbers can have perfect absorption in a narrow bandwidth, while only non-resonant techniques were known to be employed for broadband absorption. It is, of course, possible to make resonant absorbers operating in a broad range of frequencies by stacking several layers; however, this kind of broadband absorbers are very thick and bulky. In the progress of research on metamaterials (MMs); however, the resonant absorbers can have broadband behavior without aforementioned disadvantages. This classifications were made before MMs were introduced in the research field of perfect absorbers (PAs). Because this book aims to introduce metamaterial-based perfect absorber (MMPA) we will discuss geometric-transition and low-density absorbers for broadband absorbers and Salisbury screens, and Jaumann absorbers and circuit-analog absorbers (CAs) for resonant absorbers only briefly.

2.2.1 *Broadband Perfect Absorbers*

The broadband absorbers are further divided into two categories; geometric-transition absorbers and low-density absorbers. Before discussing the details of broadband PA, several important aspects should be considered. In real application of PA it is desirable to be broadband, independent of the polarization of incident EM waves, and independent of the angle of incidence. If the PA is flexible it is even better.

Consider an EM wave impinging normally on a planar boundary of medium, in the most general case, with complex permittivity and permeability. The medium is surrounded by the free space or at least air. The (complex) reflection coefficient is

$$\tilde{r} = \frac{\tilde{Z} - Z_0}{\tilde{Z} + Z_0}, \tag{2.38}$$

where $\tilde{Z} \equiv \sqrt{\frac{\tilde{\varepsilon}}{\tilde{\mu}}}$ is the impedance of the medium and $\tilde{Z}_0 \equiv \sqrt{\frac{\varepsilon_0}{\tilde{\mu}_0}} = 377\,\Omega$. For perfect reflection the absolute magnitude of impedance $|\tilde{Z}|$ is either 0 or infinity. The impedance with zero magnitude can easily be achieved by a metallic plate. Frequency-selective surface (FSS)[3] can be an example of medium with the impedance of 'effectively' infinite magnitude.

To be PA the reflection should be 0. If $\tilde{Z} = Z_0$ then $\tilde{r} = 0$. This condition is called impedance matching, at which the travelling wave cannot 'see' the reflecting boundary. To make absorption better than 99 % $|\tilde{r}|$ should be less than 0.1. One of the most important factors limiting the small enough reflection on a broad range of frequency is the dispersion of $\tilde{Z}$. In general, $\tilde{Z}$ strongly depends on frequency of incident EM wave. Therefore, the strength of dependence on frequency is very important for the performance of broadband absorption. If the impedance is very weakly dependent upon the frequency in a wide range of frequency and is close to that of free space, the reflection would be very small in a wide range of frequency.

If the angle of incidence is not zero, i.e., oblique incidence, the reflection coefficient is dependent not only on the angle of incidence but also on the polarization. From (2.25) and (2.29) we can realize that the zero reflection can be achieved at least at one angle of incidence and at a particular orientation of the electric field of incident EM wave. The reflection is usually non-zero for the other direction of polarization. It is, therefore, not a simple job to make a polarization-independent omni-directional zero reflector.

Another important factor is the material property of the absorbing medium. Even if a perfect impedance matching is achieved, the energy of the transmitted radiation should be dissipated efficiently in the interior of medium. Otherwise, the absorbed energy will be re-emitted in the form of radiation. Therefore, materials used in the broadband PA must have high electric and/or magnetic losses.

Geometric-Transition Absorbers

Geometric-transition absorber consists of two-dimensional periodic arrays of lossy-foam pyramids, cones or wedges, and is widely used in anechoic rooms to minimize the reflections from walls. Figure 2.1 shows an example of anechoic room

---

[3]Frequency-selective surface is a 2-dimensional periodic array of metallic patches, which have a particular shape for optimum operation. The screen on a microwave oven is an example. FSS is, depending on the usage, designed to reflect, transmit, or absorb EM waves based on *frequency*. The operating frequency is determined by the size of the unit cell, and the shape and the size of metallic patch.

Fig. 2.1 Photo of an anechoic room for measurements of GHz range

for measurements of GHz range. The internal appearance of EM-wave anechoic room is often very similar to that of acoustic anechoic room. Because the basic principle of absorption of acoustic wave is nearly identical to that of EM wave, the similarity of internal appearance is not surprising. The difference is the materials used in anechoic rooms. For acoustic anechoic room acoustic-wave absorbing materials are used, while radiation-absorbing ones are used for EM-wave anechoic room.

The basic principle of the operation of geometric-transition absorber is to gradually change the optical property from absorber to free space by introducing a geometrically adjusted absorbing material between them [7]. The term 'geometric-transition' is originated from this principle. In order to reduce the transmission of EM wave reflected from the rear metal plate, broadband absorbers should have sufficient attenuation. It implies that the thickness is comparable to the wavelength of incident EM waves. Although a material with permittivity, whose real part is large and loss tangent, defined as the ratio of imaginary to real parts of $\tilde{\varepsilon}$, would provide adequate loss at reduced thickness, it should be kept in mind that such a medium also have large impedance mismatch with the free space, leading to deterioration of the performance of an absorber. Therefore, the concept of tapered impedance is important to the performance of broadband absorbers.

As aforementioned, the geometrical shape can be pyramids, cones or wedges. Basically, the absorption of EM wave depends on the size of absorbing material, as well as its material properties. Therefore, geometric-transition absorber is inherently broadband one because of varying characteristic length of geometrically-shaped structures. The gradual change of the optical property suppresses reflection and results in a gradual absorption of EM wave. Furthermore, geometric-transition absorber is operating for arbitrary polarization of incident EM wave and over a wide range of angles of incidence, which is enough for the practical purposes.

When designing the geometric-transition structure, the most important factors to be considered is the material properties and the length of transition region.

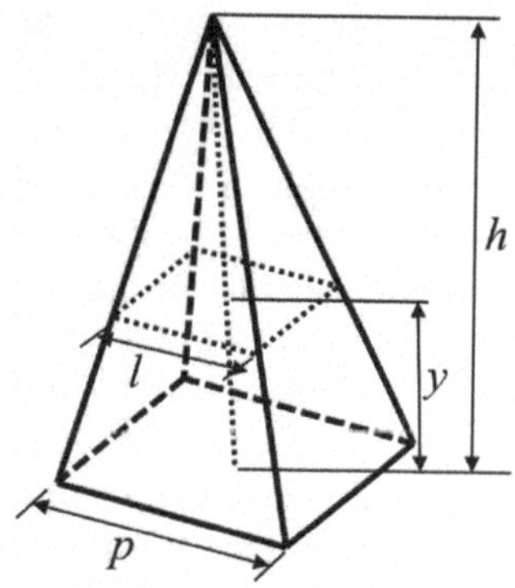

Fig. 2.2 Geometrical parameters of pyramid used in (2.39)

The length of transition should be determined by the layer thickness and should be in the order of incident wavelength. In the transition region the permittivity and the permeability should be determined by utilizing the effective-medium approximation (EMA).[4] There are two approaches to make the geometric-transition structure.

The first approach is to use absorbing medium in the form of pyramids, cones or wedges. The surface of this type of geometric-transition structures is bumpy. Let y be the distance from base to a certain height in the geometric-transition region (see Fig. 2.2). Assume that the geometric-transition region is made of a pyramid with a side p of the base and the height h defined as the distance between the central point of base and the apex of the pyramid. Pyramids with identical geometric parameters are periodically arranged in a square lattice with periodicity of p. The magnitude of a side of planar cross section of pyramid at an elevation of y is $l = py/h$. In the transition region the effective permittivity is given by [8]

$$\varepsilon_{\text{eff}}(y) = \frac{l^2}{p^2}\varepsilon_{\text{pyramid}}\frac{p^2 - l^2}{p^2}, \tag{2.39}$$

where $\varepsilon_{\text{pyramid}}$ is the bulk permittivity of material used in manufacturing pyramids. Above formula is valid only when $h \geq \frac{\lambda}{2}$ and $p \ll \lambda$, where λ is the wavelength of incident EM wave. A similar approach can be applied for the case of cone- or wedge-type transition.

The second approach is laminated absorbers. Because of lamination the surface of this type of absorbers is flat. The basic principle of laminated absorbers is nearly identical to that of broadband anti-reflective (AR) coating which utilizes multilayer interference by stacking homogeneous layers with different refractive indices and thicknesses. The choice of refractive indices and thicknesses depends on the desired performance. In a laminated absorber planar layers of homogeneous media with different permittivity are prepared with flexible adhesives and latex. Consequently, they are flexible and even able to be wrapped. The arrangement of multiple layers is

[4]See Sect. 2.3 for the detailed discussion of EMA.

made in such a way that the permittivity or the impedance is decreased from the bottom to the top. The change of permittivity is not gradual but discontinuous.

Combination of the two approaches is also possible. Absorbers utilizing the idea of this combination is called hybrid absorbers. The frequency range of pyramidal absorber depends on the size of the geometric-transition structure. Since the size of pyramids should be large for low-frequency performance, the size of whole absorbers also become very large. Therefore, for instance, in a compact anechoic room, the pyramidal absorber has inherently poor performance in low-frequency. To improve the low-frequency performance Ford and Chambers [8] put one or more FSSs underneath the pyramidal absorber. They observed a significant improvement of the low-frequency absorption bandwidth, when either a bilayer of single-loop FSSs or a single layer of double-loop FSS is embedded within the base region of a pyramidal absorber, over those of both the unmodified absorber and one with a single-loop FSS loaded. Later, Holtby et al. [9] further modified this structure by replacing the bilayer of single-loop FSSs or single layer of double-loop FSS with binary FSS, and observed a very significant improvement of low-frequency absorption. They used the genetic algorithm [10] for optimization of the binary FSS.

Another important factor to be considered is the choice of absorbing material and their shape. Neither perfect conductor nor perfect insulator can be used for the absorbing material, since a perfect conductor completely reflects incident radiation and a perfect insulator is completely transparent owing to the absence of the imaginary part of permittivity. Ferrite-tile, broadband-foam, laminated and hybrid absorbers are commercially available [11]. The operational principle of ferrite-tile absorbers is based on a particular material property of certain ferrites with

$$\frac{\tilde{\varepsilon}}{\varepsilon_0} = \frac{\tilde{\mu}}{\mu_0}. \tag{2.40}$$

Therefore, the impedance of these ferrites are that of the free space. As a consequence, their surface reflects no incident EM waves. Once EM waves hits the surface of ferrite-tile absorber, it transmits completely, and the transmitted EM wave will be absorbed by the ferrite, which has not only high loss tangent but also large imaginary part of permeability. Therefore, both electric and magnetic absorption occur. Ferrite-tile absorbers are not exactly 'geometric-transition' absorbers, because the permittivity does not change gradually, but abruptly. Ferrite-tile absorbers; however, can be classified as geometric-transition absorbers with zero transition region.

Low-density absorbers

Low-density absorbers for acoustic noise reduction utilize very porous or sparse material. On the ceiling of classrooms, business offices, and other rooms requiring noise reduction this type of absorbers is easily found. They are usually made of porous open cell foams, typically open-cell rubber foams or melamine sponges, which are highly effective for absorbing noises from medium- to high-frequency

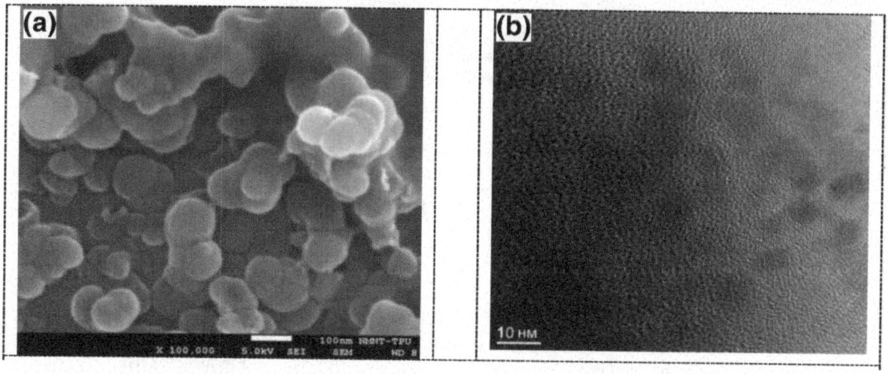

Fig. 2.3 Electron-microscopy image of nanoglobules of the inter-porous barrier of the foam material: **a** scanning electron microscopy and **b** high-resolution transmission-electron microscopy. Reproduced from [12] with permission

ranges, but not for low-frequency one. The basic principle of low-density acoustic absorber is dissipation of the energy of noise by thermal conduction and viscous friction inside the cell structure. Low-density absorbers of EM waves with porous materials operate in a very similar way of porous acoustic absorber. Once EM waves enter a pore, it hit the surface of absorbing material several times before it eventually leave the pore. The multiple reflections at the interior surfaces of the pore enhance the absorption because a part of EM wave is absorbed whenever it hit the surface of absorbing material. The multiple reflections also effectively increase the path length, leading to a reduced size of absorbers.

The effects of pores are investigated by, for example, Kazmina et al. [12] who investigated the physical and mechanical properties of a foam glass-ceramic material synthesized from raw siliceous-earth material by the two-stage method at temperatures below 950 °C, and observed improvement of its physical and mechanical properties in comparison with foam glass synthesized from glass cullet. As shown in Fig. 2.3, inter-porous barrier, nanosized structural elements, are observed that can be represented as quartz or cristobalite particles on the surface of which one-dimensional Si–O chains are localized. The effects of pores are evident in Fig. 2.4 in which the measured absorption is significantly enhanced in the sample synthesized from the foam glass-ceramic material compared to that of the sample synthesized from commercially-available glass cullet. The enhanced absorption was attributed to the interaction of an EM field with carbon nanoparticles remained after foaming due to incomplete transition of the carbon froth to the gas phase. Li et al. [13] prepared porous carbon fibers and carbon nanofibers from polyacrylonitrile/polymethylmethacrylate blend fibers to compare the absorption performances. They found that composites filled with the porous carbon fibers exhibited a much better performance in microwave absorption than those containing the carbon nanofibers. The enhanced absorption in the porous carbon fibers is due to the combination of

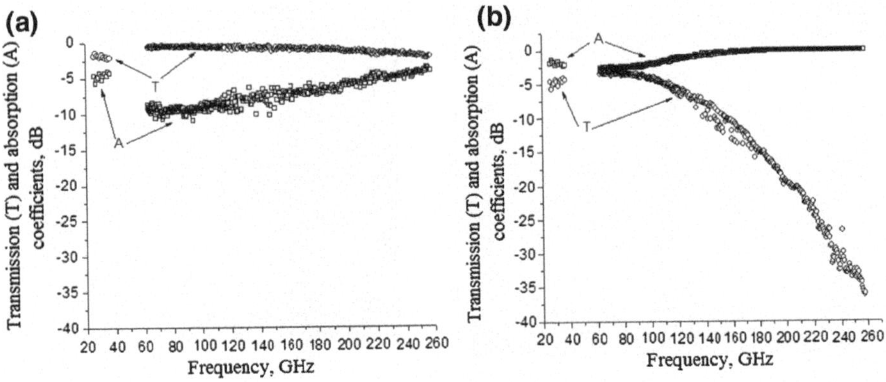

Fig. 2.4 Parameters of the EM response of the sample synthesized from commercially-available glass cullet (*left*) and synthesized from the foam glass-ceramic material (*right*). Reproduced from [12] with permission

the dielectric-type absorption and the interference of multiple reflections of microwaves.

Sparse material for broadband absorbers can be made by dispersing small absorbent particles in a matrix material. Lee et al. [14] have measured the complex permittivity spectra and the conductivity of graphite nanoplatelet (GNP)/epoxy composites in a range of 0.05–18.0 GHz. They made two different samples; one with as-received GNPs and the other with nitric acid-treated GNPs. Although the as-received GNPs are already treated with sulfuric acid and rapidly heated to 900 °C for expansion, further acid treatment with a 67 % nitric-acid solution improves the absorption significantly (see Fig. 2.5). Both real and imaginary parts of the permittivity are increased by the additional nitric-acid treatment. The enhanced performance was ascribed to the increased functional groups such as carboxylic-acid groups on the GNP surface, which can improve interfacial bonding to the surrounding epoxy matrix. The authors suggested that this composite material can be used as conductive coatings or microwave absorbing materials with high structural integrity.

EM waves consist of spatio-temporally varying electric and magnetic fields. Therefore, it is possible to absorb energy of incident wave magnetically, as well as electrically. In dielectrics or conductors the absorption of EM wave is due to the dielectric loss caused by the loss tangent for dielectric or the Ohmic loss caused by the conductivity of conductors. The loss tangent and the conductivity are effectively the imaginary part of permittivity. In magnetic materials changes in the alignment and agitation of the magnetization lead to losses. Ji et al. [15] tried an idea to incorporate both electric and magnetic media together in a aliphatic polyurethane resin in which the incorporated electric and magnetic losses might improve the absorption. Carbonyl iron particles and flake graphite (FG) were used for the magnetic absorber and the electric absorber, respectively. They found that the added

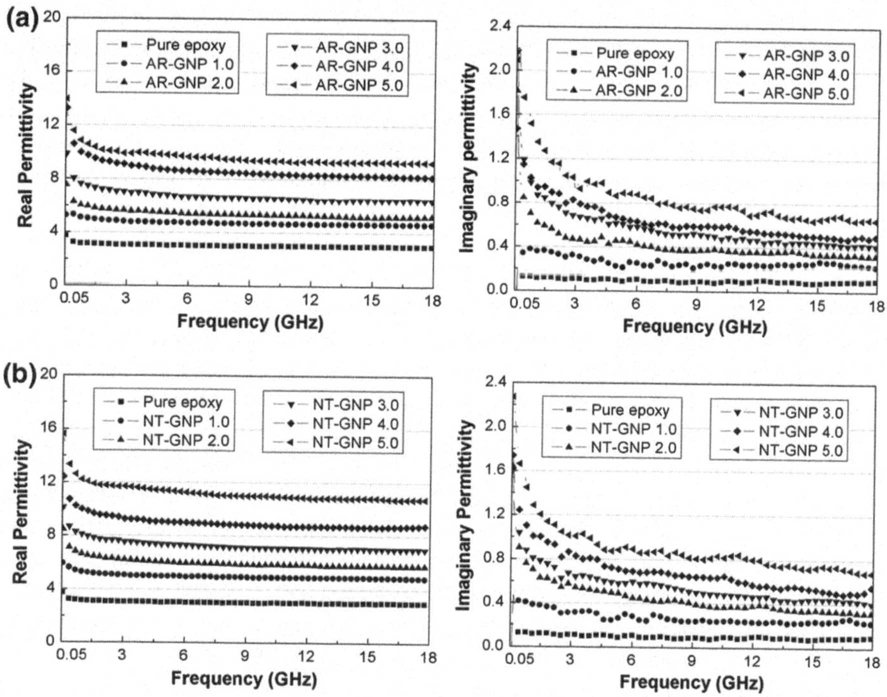

Fig. 2.5 Real (*left panels*) and imaginary (*right panels*) parts of the permittivity of epoxy composites with **a** as-received GNPs and **b** nitric acid-treated GNPs. Reproduced from [14] with permission

FG can effectively regulate the EM parameters of the hybrid, resulting in the increasing of dielectric loss.

2.2.2 *Resonant Perfect Absorbers*

Resonant EM-wave absorbers utilize the quarter-wavelength thickness of the substrate before the grounded metallic plate. The first resonant EM-wave absorber was invented by American engineer Salisbury [16], and it was called Salisbury screens [17] named after the inventor. The basic principle of the operation of Salisbury screens is exactly the same as that of the AR coatings. A Salisbury screen consists of three layers; (i) a thin resistive screen, (ii) a low-loss dielectric substrate with an exact thickness corresponding to a quarter of the wavelength of incident EM wave, and (iii) a continuous metallic surface. If the incident EM wave pass through the first layer, a thin resistive screen, is reflected by the continuous metallic surface, and eventually reach the resistive screen again, then the distance traveled is exactly a half wavelength, which corresponds to 180° phase change. Therefore, there is an exactly 180° phase difference between the EM wave just incident upon the first layer and the EM waves reflected from the continuous metallic surface, resulting in a destructive interference and, consequently, no reflection. This is exactly the operating principle of the AR coating. The Salisbury screen; however, has two major drawbacks; the thickness and the absorption band. To satisfy the complete destructive interference, the dielectric layer has a thickness of quarter wavelength divided by its refractive index. If the wavelength of the incident EM wave is very long then the dielectric layer should be very thick. Because for the Salisbury screen the absorption occurs only at a specific wavelength satisfying the condition for complete destructive interference, the absorption bandwidth is very narrow. These two drawbacks hinder the Salisbury screen to be used in practical applications.

Another way to understand the principle of operation of Salisbury screen is its equivalent circuit; however, we need to know the transmission-line theory (TLT).[5] The quarter-wavelength transmission line transforms the short circuit at the rear metal plate into an open circuit at the resistive screen. Therefore, the effective input impedance of the short-circuited transmission line becomes infinity because the load impedance is zero owing to the short-circuited, rear metal plate (for more detail, refer to Sect. 2.4). It implies that the incident EM wave can 'see' the resistive screen only. As a consequence, if the resistive screen has an impedance identical to that of the free space $Z_0 = 377\,\Omega$, there is no reflection at resistive screen. If the wavelength of incident wave is longer than 4 times the distance between the resistive screen and rear metal plate, the impedance of the transmission line becomes inductive and the reflection of incident EM wave occurs.

In the original design of Salisbury screen there was an air gap between the resistive screen and rear metal plate. If the air gap is replaced by a high-permittivity, low-loss dielectric substrate we can reduce the thickness of the Salisbury screen. The dielectric substrate has a low input impedance of the transmission line, resulting in a narrowed bandwidth. Although ferrites or other materials with a higher intrinsic impedance than that of the free space can be used as materials for the middle layer, it is not desirable to use them because they are usually very heavy [17].

One of the two major drawbacks of Salisbury screen can be relaxed. The narrow bandwidth can be extended by putting one more layers of resistive screen and quarter-wavelength gap in front of Salisbury screen. This modified Salisbury screen is called Jaumann absorber because it was first realized by J. Jaumann during World War II in Germany. The historical background of Jaumann absorber is nearly unknown, probably because it was invented during World War II in Germany. Severin [18] was the first one who calculated the reflection spectrum of Jaumann absorber. By putting one more layer of resistive screen and quarter-wavelength gap in front of Salisbury screen with the aid of TLT he was able to calculate the reflection spectrum, which has dual minima with zero reflectivity, of Jaumann absorber. He even calculated multilayered Jaumann absorber in which several

[5]The transmission-line theory will be discussed in Sect. 2.4.

layers of resistive sheets and quarter-wavelength gaps were put in front of Salisbury screen. There are several reflection minima in the reflection spectrum. This is exactly the same strategy as that of the AR coatings with broadband operation. In many senses, Jaumann absorber can be considered as a simple extension of Salisbury screen.

Another kind of Salisbury screen is Dällenbach-layer absorber (DLA) [19] in which only a homogeneous single layer was placed in front of grounded metallic plane. The thickness, the permittivity and the permeability are adjusted in such a way that the reflection is minimized. According to TLT, thc input impedance should satisfy the relationship

$$\tilde{Z}_i = \tilde{Z}_c \frac{\tilde{Z}_L + \tilde{Z}_c \tanh(\tilde{k}l)}{\tilde{Z}_c + \tilde{Z}_L \tanh(\tilde{k}l)}, \tag{2.41}$$

where l and $\tilde{k}$ is the length and the complex wave number, respectively, of the transmission line, and $\tilde{Z}_c$ and $\tilde{Z}_L$ is the characteristic and load impedances, respectively. For DLA the load impedance is zero because the metal plate is effectively short circuit, and the input impedance is that of the free space. Therefore,

$$Z_0 = \tilde{Z}_c \tanh(\tilde{k}l). \tag{2.42}$$

Because the left-hand side is real, the sum of the phase angles of Z_c and $\tanh(\tilde{k}l)$ should be zero. It implies that there is destructive interference at the boundary. Therefore, the thickness of the absorbing layer is quarter wavelength. In the design of DLA not only the thickness of the absorbing layer but also its $\tilde{\varepsilon}$ and $\tilde{\mu}$ should be adjusted carefully. As in the case of Salisbury screens, it is not possible to obtain a broadband behavior in DLA with a single absorbing layer. As in the case of Jaumann absorbers; however, by stacking multiple layers with different absorption bands, a broadband behavior can be obtained.

In a sense, both Salisbury screens and its extension, Jaumann absorbers, are actually a kind of the so-called CAs. In Salisbury screens and Jaumann absorbers only one or more resistive sheets are used, and thus their impedances do not contain any imaginary components. The basic structure of CAs is identical to that of Salisbury screens or Jaumann absorbers; tri-layer structure with quarter-wavelength thickness for the middle layer. By putting a layer of periodically-arranged conducting patches instead of continuous resistive sheet the front layer becomes reactive, as well as resistive. The shape of elemental patch could be continuous-wire type, cut-wire (CW) type, cross type, or other types. The resistance of patch itself can be modelled as a resistor. The thickness of the patch is important to adjust the resistance. In addition, the linear part of patch plays a role of inductor and the gap between any two adjacent patches plays a role of capacitor. Therefore, the transmission line of CAs can be considered as a one- or two-dimensional infinite array of identical *RLC* circuits, which are connected in series. The resonance frequency is selected by the resonance of *RLC* circuits, not by the distance between first and third layers. It is the reason why this type of absorbers is called 'circuit-analog' absorbers. Because the resonance frequency is determined by the frontal surface, we call the frontal surface made of a periodically-patterned metallic patches a 'frequency-selective' surface. The term 'circuit analog' is derived from the fact that the geometrical patterns of lossy patches can be modelled as an equivalent 'circuit' with their effective resistance, capacitance, and inductance. As a result, the equivalent-circuit theory can be used in the conceptual analysis and design of the resultant absorber, although it can not give us a quantitatively accurate result.

The role of middle layer (quarter-wavelength gap) is not as crucial as in the case of Salisbury screens or Jaumann absorbers. In the later, the thickness of the middle layer not only determines the resonance frequency but also minimize the reflection by destructive interference. In CAs the middle layer also plays a similar role. The thickness of the middle layer of CAs is, of course, approximately chosen to be the quarter wavelength of the resonance frequency of the *RLC* circuit of FSS. The behaviors of CAs are different from those of Salisbury screens or Jaumann absorbers if the incident EM wave has off-resonance frequency. As is discussed above, if the frequency of incident EM wave is lower than the resonance frequency, the impedance of middle layer becomes inductive and there is a significant reflection in case of frequency far from the resonance. In case of CAs the input impedance toward the rear metal plate also becomes inductive for the low-frequency EM wave, while that of FSS becomes capacitive. Therefore, the admittance, which is defined as the inverse of impedance, of the whole CA is the sum of the admittance toward the ground metal plate and that of FSS. The effects of these two components are the cancelation of reactive parts of both admittances, leading to very small reflectance. For frequencies higher than the center frequency (resonance frequency of FSS) the roles of the middle layer and FSS are interchanged, but the effects are the same as the case of low frequency. It implies that CAs have broader bandwidth than Salisbury screens or Jaumann absorbers. As in the case of Jaumann absorbers, CAs can be with multiple layers of CA which is modified appropriately to achieve the maximum bandwidth.

In actual application absorbers should possess a broadband response and should be independent of the polarization of incident wave and of the angle of incidence. CAs partially solve the problem of narrow bandwidth of Salisbury screen. CAs can also be polarization independent by making pattern of patches in a symmetric shape, such as square, hexagonal, octagonal or circular shapes. The independence on the angle of incidence; however, seems to be inherently impossible to achieve. Two problems can be encountered. The first one is that the resonance frequency is determined by the distance travelled by the incident EM wave inside the second layer. If the EM wave is obliquely incident, the distance travelled is longer than that of the normal incidence, resulting in a blue-shift of resonance frequency. For the quarter-wavelength thickness, i.e., $nd = \lambda/4$, there is destructive interference for normal incidence if the incident EM wave has exactly the wavelength satisfying this condition. For oblique incidence the optical-path difference, $2nd\cos\theta_t$, is shorter than that for the normal incidence, $2nd$. It implies that the complete destructive interference occurs for the EM wave with shorter wavelength than in case of the normal incidence, resulting in a blue-shift of the central peak in absorption spectrum. The second one is that the impedance-matching condition for total transmission at the first interface is dependent upon the angle of incidence. Munk et al. [20] tried to attack the problems by inserting a stratified dielectric material and/or a dielectric-matching slab between the resistive sheet and metallic plate in Salisbury-screen CAs. Although the results were not complete solution for the problems, they were successful to obtain 1.3–1.4 times enhanced bandwidth and partially release the restrictions for the incident angle.

2.2.3 Metamaterial-Based Perfect Absorbers

MMPA is in a sense a kind of resonant PAs, more specifically CAs. Most of MMPAs consist of three layers; (i) a layer of periodically-arranged metallic patterns, (ii) a dielectric layer and (iii) a continuous metallic layer. It is nearly the same as the case of CA. MMPAs use a layer of periodically-arranged metallic patterns, which is similar to the FSS of CAs; however, the thickness of the dielectric layer could be much smaller than the wavelength, especially, in the GHz regime. To accomplish the total absorption there should be no reflection and no transmission. No transmission can be accomplished by the third layer, a continuous metallic plate, which completely blocks all incident EM waves. Since the third metallic layer simply reflects all EM waves falling upon it, its role is simply providing zero impedance (or short circuit) for the interpretation of the experimental results by using TLT.

The basic role of dielectric layer is to provide a space to the incident EM wave to stay and be absorbed. It is generally believed that, in order for the dielectric layer to provide the incident EM wave with a sufficient space, it is desirable to use high-ε_1 material because it reduces the thickness of dielectric layer with the optical path still maintained. For the incident EM wave, the path should be given by nd, where n is the refractive index and d is the distance traveled. Although it is generally true, it is not necessary for MMPA to be very thick, satisfying, for instance, the quarter-wavelength condition. Unlike MMPAs, the quarter-wavelength condition is crucial for the operation of AR coating and Salisbury screen. Dielectric materials usually have very small imaginary part of permittivity, and this small imaginary part is sometimes enough to absorb all EM waves incident during travelling inside the dielectric. The imaginary part, ε_2, of dielectric layer in MMPAs plays a different role from that of Salisbury screen. In Salisbury screen, the dielectric layer is supposed to be lossless, i.e., zero loss tangent to satisfy the complete destructive interference, implying that $\varepsilon_2 = 0$. In MMPAs; however, the imaginary part is important for the absorption of the incident EM waves. In addition to providing a space for the incident EM wave to be absorbed, the dielectric layer in MMPAs plays another role, and the role depends on the specific design. Sometimes it plays a role of, for instance, the Fabry-Pérot cavity [21], which is similar to the case of Salisbury screen. One should notice that, even though the role of dielectric layer is the Fabry-Pérot cavity in [21], it does not only provide the enough space for the cavity resonance but also it produces another absorption, resulting in a dual-band absorption.

There are two absorption mechanisms in MMPA; (i) the dielectric loss and (ii) the Ohmic loss. The dielectric loss is the dissipation of energy of EM wave occurring in the interior of a dielectric medium. The loss tangent determines the degree of loss. As a result of the causality and the non-locality of time, any material should be simultaneously dispersive and absorptive to any EM wave. Material, of course, can be lossless in some finite frequency ranges and dispersionless in other finite frequency ranges; however, it is not possible to have a frequency range in which the material is both lossless and dispersionless. It means that any *good* insulator should have even though very small, but non-zero dissipation of energy of EM waves. See Sect. 2.1.2. Ohmic loss is the energy dissipation in an electrical circuit arising from circuit resistance when current flows. The both losses usually lead to a heating-up of electric circuits. We call it Joule heating.

Because MMPAs are usually composed of both dielectric and metallic materials, it is possible to have either of two loss mechanisms. Sometimes, even both mechanisms play comparable roles. In the first MMPA the dominant absorption was due to the dielectric loss. Landy et al. [22] theoretically and experimentally reported that MMPA in the GHz range, which was composed of conducting electric resonators on both front and rear sides of a dielectric substrate. The experimentally-observed absorption peak is at 11.5 GHz with an absorption of 88 %, while in the simulation it would be at 11.48 GHz with an absorption of 96 %. They attributed the discrepancy to the imperfection of the patterns introduced during fabrication. They also theoretically investigated the origin of the absorption mechanism, and found that a very weak energy dissipation occurred on the metallic surface of the first layer, while the dissipation was very strong in the interior of dielectric layer. Consequently, it was concluded that the dielectric loss occurred in between two MM elements where the electric field was large.

Hu et al. [23] attributed the discrepancy between experiment and simulation to the considerable influence of the dielectric loss on the strength of electric and magnetic resonances. They designed and fabricated MMPA with unit cell made of a CW on top and a split-CW at the bottom, which was analogous to the conventional CWP. In experiment an absorption peak appeared at 4.87 GHz with an absorption of 94.84 %. In simulation the peak appeared at the same frequency as the experiment; however, the absorption was slightly better than the experiment. They found that the discrepancy was due to the fact that ε_2 had considerable influence on the resonant system in MM absorber, not only behaving as a source to introduce loss, but also playing a role of an effective element which could be utilized to tuning the strength of electric and magnetic resonances.

Very recently Li et al. [24] theoretically investigated the possibility of MMPA without the patterned front metallic layer to enhance the bandwidth. The dielectric E-shaped resonators, which were made of $SrTiO_3$ with $\varepsilon_2 = 300$ and loss tangent of 0.005, were placed on the copper plate. They argued that the dielectric MMPA based on dielectric resonances could provide modulated effective permittivity and permeability to meet the band-enhanced impedance matching and the strong EM attenuation simultaneously.

It is generally believed that the Ohmic loss in MMPA is very weak in the optical range because metals are very lossy owing to the interband transitions, which absorb light quanta with the energy corresponding to the energy difference between the occupied level and the empty level into which the electrons from the occupied level eventually transits. However, in the lower frequency, especially in the GHz range, metals usually very good conductor with little loss. The estimated fraction of Ohmic loss in a THz perfect transmitter is only 0.1 % or less [25]. The estimation was done by assuming that in the THz range the permittivity of most of metals obey the Drude model and loss fraction is of the order of the ratio of the skin depth to the wavelength. As frequency increases; however, the Ohmic loss starts to dominate over the dielectric loss. Especially, from the infrared (IR) and higher-frequency range the dielectric loss becomes weaker and weaker, and eventually it can be neglected.

Although in the GHz range the role of Ohmic loss is supposed to be negligible, it is not impossible to induce the Ohmic loss in the GHz range. Tuong et al. [26] theoretically investigated a possibility to control the absorption mechanism. To enhance the Ohmic loss the thickness of copper square in the frontal patterned metallic layer was adjusted to obtain the absorption peak. By controlling the sheet resistance of the copper square it was possible to obtain nearly unity absorption at 16 GHz. By adjusting the dielectric loss tangent it was possible to obtain the near perfect absorption at the same frequency. As can be seen in Fig. 2.6, it is possible to control the loss mechanism. Another interesting aspect of the investigated structure is its independence of absorption on the polarization of incident EM wave due to the symmetrical shape of the metallic patches.

Furthermore, the Ohmic loss leads to an undesired heat in the photovoltaic devices. It was reported; however, the undesired heat generated by the Ohmic loss can be effectively converted into photovoltaic currents by the exchange of

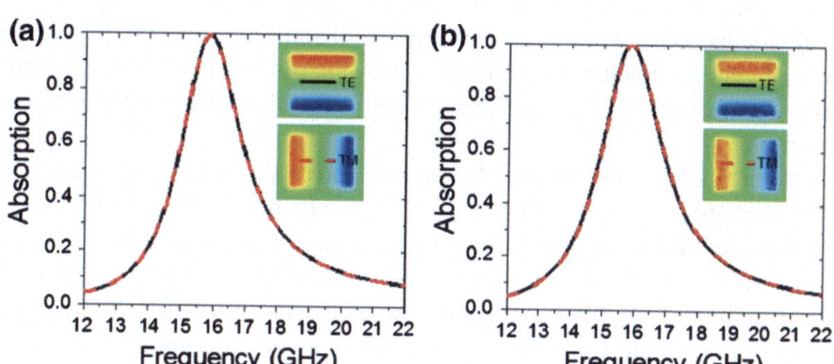

Fig. 2.6 Compared spectra between TE and TM polarizations for the **a** Ohmic-loss and **b** dielectric-loss PA peaks. The *insets* are respective induced fields (*Ez*) in the xy-cutting plane. Reproduced from [26] with permission

undesired resistive losses with the useful optical absorbance in the active semiconductors. Vora et al. [27] was able to figure out how to convert the resistive heat into about 75–95 % photovoltaic absorption by designing a MMPA with $In_xGa_{1-x}N$ spacers.

If all incident EM waves onto a surface are transmitted there is no reflection. This perfect transmission is essential for MMPA. To achieve perfect transmission the surface should have the impedance matched to that of the free space. It can be thought that the role of the first layer is very similar to that of the resistive screen of Salisbury screen; however, there are a few very important differences.

First, the transmission at the first layer is led by a certain resonance in the targeted frequency regime, unlike the case of Salisbury screen. According to the Drude model, the real part of permittivity, ε_1, of most metals in the IR and lower-frequency ranges can be approximately represented as

$$\varepsilon_1(\omega) = \varepsilon_\infty - \varepsilon_0 \frac{\omega_p^2 \tau^2}{1 + \omega^2 \tau^2}, \qquad (2.43)$$

where ω and τ are the angular frequency of the incident EM wave and the relaxation time, respectively. If $\omega\tau \gg 1$, (2.43) is further approximated as

$$\varepsilon_1(\omega) = \varepsilon_\infty - \varepsilon_0 \frac{\omega_p^2}{\omega^2}. \qquad (2.44)$$

ε_∞ can be obtained as follows. ε_1 in the region satisfying the condition $\omega\tau \gg 1$ can be fitted as a function of ω^{-2} or λ^2. Equation (2.44) implies that the plot ε_1-versus-ω^{-2} or ε_1-versus-λ^2 is a straight line and ε_∞ is the intercept at $\omega^{-2} = 0$ or $\lambda^2 = 0$. ε_∞ is usually different from ε_0, the permittivity of vacuum, since it is not $\varepsilon_1(\infty)$, which is the real part of permittivity measured at $\omega = \infty$ and is ε_0. It implies that if we measure ε_1 in a very high-frequency region, such as gamma-ray region, it is ε_0, since most materials are transparent in this region. The intercept of experimental ε_1-versus-ω^{-2} plot of a metal in the IR and the lower-frequency ranges is not ε_0. Instead, it is some other value, ε_∞, since in the IR and the lower-frequency ranges metals are in general opaque. In addition to this, the polarization of core electrons of metal ions comes into play. ε_∞ represents the integration of contributions from all core electrons.

Most of metals have the plasma frequency corresponding to the energy greater than 5 eV (see, for example, Ordal et al. [28]), which is in the ultraviolet region. Therefore, ε_1 is negative with very large absolute value in the IR and the lower-frequency regimes. If one wants to fabricate MMs operating in the GHz, the THz, the IR and even visible regimes it is necessary to lower the 'effective' plasma frequency to the targeted frequency regime, since the plasma oscillations is usually crucial for attaining the desired properties revealed in the targeted frequency regime.

Because the plasma frequency is proportional to the square root of the charge density [see (2.37)], one needs to make 'diluted' charge density, since only the charge density can be tuned by utilizing a proper design of metallic patterns and the other quantities (e, m^* and ε_0) are not possible to alter. Here, the term 'diluted' implies that the 'effective' charge density in MMs should be properly *diluted* to facilitate the plasmonic resonance in the targeted frequency regime. It is the reason that an array of the patterned unit cells is necessary for MMs. By partially removing the metallic part, which was originally continuous, the 'effective' charge density can be properly adjusted to facilitate the plasmonic resonance in the targeted frequency regime.

Second, the first layer of MMPA plays an active role, while the role of resistive screen of the Salisbury screen is passive. The thin resistive screen of the Salisbury screen partially reflects the incident EM wave back to the space, where the incident wave comes, and partially transmits the reflected wave from a continuous metallic surface. Eventually, these two EM waves interfere destructively, leading to no reflection. No reflection would be accomplished in MMPAs not by destructive interference, but by requiring that the impedance of the first layer should match to that of vacuum or air. The incident EM wave cannot notice the difference between the first layer and the atmosphere surrounding it because of impedance matching. The perfect impedance matching implies that the two media, the atmosphere and the first layer, are optically identical. Ordinary natural materials are nearly impossible to possess the impedance perfectly matched with that of environment. Only MMs, whose geometric parameters of metallic patterns are carefully adjusted, can possess such properties with ease. Such impedance matching can only be achieved through certain resonances or plasmonic couplings, as argued in the previous paragraph. Therefore, MMs should be treated as a *homogeneous* materials even though it is macroscopically inhomogeneous. As a consequence, EMA is necessary for the analysis of MMPA.

The first layer made of a periodic array of metallic patches should be designed carefully to satisfy the impedance-matching condition with its environment. A brief discussion of the design strategy of MMPA can be found in [29].

2.3 Effective-Medium Approximation

Usually, MMs are inevitably fabricated as composite, i.e., *inhomogeneous* structures. As aforementioned; however, the first layer, composed of intentionally-designed metallic patterns with a periodic arrangement, should be treated as *homogeneous* to accomplish the perfect transmission of incident EM waves. This is the reason why EMA, or often called as 'effective-medium theory' is especially necessary for the first layer (MM layer) of MMPA.

In the microscopic level the classical mechanics is no longer valid, and quantum mechanics comes into play. Although all of the macroscopic material properties are determined in principle by the details of wave functions of particles, more specifically, the electronic structures of material, in quantum regime, individual quantum-mechanical effects are not easily detectable in macroscopic-level measurements. Furthermore, when EM wave, whose frequency is lower than X-ray range, is interacting with materials, the wavelength of EM wave is usually much longer than the size of constituent atoms and molecules. Therefore, the EM wave cannot 'see' the atomic-level details of electronic structures of materials, which determine the material properties observed in the macroscopic level. Therefore, the material can be treated as *homogeneous*. It is the background of EMA. This background can be extended to the macroscopic level unless the inhomogeneity of composite materials plays a significant role. It is, therefore, generally believed that the long-wavelength limit is crucial for the application of EMA. We will introduce two mostly used EMAs; Bruggeman's and Maxwell-Garnett's.

Consider a system consisting of a medium whose permittivity is ε_s. The permittivity should satisfy the relationship

$$\frac{\varepsilon_s - \varepsilon_0}{\varepsilon_s + 2\varepsilon_0} = \frac{\mathcal{N}}{3\varepsilon_0}\alpha, \tag{2.45}$$

where $\mathcal{N}$ is the number of polarizable molecules per unit volume and α is the molecular polarizability. If the polarizable entity is in spherical shape, from the elementary electrostatics, we can get

$$\alpha = 4\pi r^3 \varepsilon_0 \frac{\varepsilon_s - \varepsilon_0}{\varepsilon_s + 2\varepsilon_0}, \tag{2.46}$$

where r is the radius of spherical entity. If this spherical entity is embedded in a dielectric medium of permittivity ε_m then the effective permittivity (ε_{eff}) satisfies the relationship

$$\frac{\varepsilon_{\text{eff}} - \varepsilon_m}{\varepsilon_{\text{eff}} + 2\varepsilon_m} = f\frac{\varepsilon_s - \varepsilon_m}{\varepsilon_s + 2\varepsilon_m}, \tag{2.47}$$

where f is the volume fraction of the spherical entity. This is the famous Clausius-Mossotti equation. In the Bruggeman EMA it is assumed that a system consists of spherical inclusions with permittivity of ε_s, which are homogeneously dispersed in the host medium with permittivity of ε_h. The effective permittivity of composite medium can be obtained by the condition

$$f\frac{\varepsilon_{\text{eff}} - \varepsilon_s}{\varepsilon_{\text{eff}} + 2\varepsilon_s} + (1-f)\frac{\varepsilon_{\text{eff}} - \varepsilon_h}{\varepsilon_{\text{eff}} + 2\varepsilon_h} = 0. \tag{2.48}$$

This relationship can be generalize easily to a multi-components system as [30]

$$\sum_{i=1}^{n} f_i \frac{\varepsilon_i - \varepsilon_{eff}}{\varepsilon_i + 2\varepsilon_{eff}} = 0, \tag{2.49}$$

where f_i and ε_i are the volume fraction and the permittivity of ith medium, respectively, and ε_{eff} is the *effective* permittivity of composite medium. The volume fractions should satisfy the condition, $\sum_i f_i = 1$. It is tempting to extend the Clausius-Mossotti equation, (2.47) to a multi-components system as

$$\frac{\varepsilon_{\text{eff}} - \varepsilon_0}{\varepsilon_{\text{eff}} + 2\varepsilon_0} = \sum_i f_i \frac{\varepsilon_i - \varepsilon_0}{\varepsilon_i + 2\varepsilon_0}. \tag{2.50}$$

Here, the system is assume to be immersed in vacuum. Such extension of the Clausius-Mossotti equation is not generally possible. A different extension; however, is possible by using the Lorentz local-field concept as

$$\frac{\varepsilon_{\text{eff}} - \varepsilon_0}{\varepsilon_{\text{eff}} + 2\varepsilon_0} = \frac{1}{3\varepsilon_0} \sum_k \mathcal{N}_k \alpha_k, \tag{2.51}$$

where $\mathcal{N}_k$ and α_k are the number density and the molecular polarizability of kth species. A detailed derivation of Bruggeman EMA and its historical backgrounds can be found in [31].

In the Maxwell-Garnett EMA the effective permittivity of composite medium made of spherical entities, which are homogeneously dispersed in a host medium, with the permittivity of ε_s can be directly obtained by using the multi-components Clausius-Mossotti relation (2.51). By combining (2.46) and (2.51) and using

$$f = \frac{4\pi}{3} r^3 \mathcal{N} \tag{2.52}$$

we can get ε_{eff} which satisfies the relationship [32]

$$\frac{\varepsilon_{eff} - \varepsilon_h}{\varepsilon_{eff} + 2\varepsilon_h} = f_s \frac{\varepsilon_s - \varepsilon_h}{\varepsilon_s + 2\varepsilon_h}, \tag{2.53}$$

where ε_h is the permittivity of host medium. Then ε_{eff} can be solved as

$$\varepsilon_{eff} = \varepsilon_h \frac{2(1-f)\varepsilon_h + (1+2f)\varepsilon_s}{(2+f)\varepsilon_h + (1-f)\varepsilon_s}. \tag{2.54}$$

The Bruggeman EMA has some advantages over the Maxwell-Garnett EMA, since the Bruggeman EMA treats all inclusions and the matrix on an equal footing. In the Maxwell-Garnett model the inclusions are supposed to have small volume fractions because the inclusions are assumed to be spatially separated. For more detailed discussion and generalization see [33].

EMA can be used in various ways. It is especially powerful when only ε_{eff} is experimentally accessible and the other information, such as volume fractions, permittivities of composite media and so on, is limited. Rhee et al. [34] applied the Bruggeman EMA to the measured optical conductivities of Ce thin films deposited under various conditions in order to find the optimum spectrum of optical conductivity without any surface roughness. Because the films were deposited under various conditions the surface roughness also varied. The 'effective' optical conductivities of various samples were measured by a rotating-analyzer ellipsometer. The rough surface was modelled as a combination of voids and Ce by using the Bruggeman EMA. By applying the three-phase model, air/rough-layer/bulk-Ce, the effects of surface roughness were corrected and the optimum values of optical conductivity were obtained.

Recently, a skepticism on the validity of long-wavelength limit on EMA has been casted. It is generally accepted that the size of the unit cell of MM should be a few times smaller than the wavelength of incident EM wave. This is the so-called deep-subwavelength condition. The actual size of the unit cell of MM; however, is sometimes comparable to the wavelength of incident EM wave (see, for example, Resink et al. [35]). It implies that the necessity of the deep-subwavelength condition for the EMA can be a subject to the dispute. Furthermore, an EMA beyond the long-wavelength limit has been developed by Wu et al. [36]. To verify the validity of the theory, Wu et al. performed band-structure calculations on a periodic composite in two dimensions for a specific configuration with complicated Mie resonances in the microstructure. The theory can be valid even when the wavelength of incident EM wave can be as large as 1.3 times the lattice constant, while the long-wavelength theory breaks down when the wavelength is 3–4 times smaller than the lattice constant.

Very recently, a generalized EMA for MMs was proposed [37]. The authors tested the validity of the long-wavelength limit and concluded that the theory does not require any additional long-wavelength approximation. Essentially the same approach made by Wu et al. [36] was applied, and it as predicted that the refractive index effectively becomes infinitely large as the frequency of incident EM wave approaches the resonance. It means that the wavelength becomes very short near the resonance, therefore, the long-wavelength approximation cannot be applied. The model system consists of spherical magneto-dielectric inclusions in a matrix. The spherical inclusion has a core-shell structure, in which the incident EM wave does not scatter. In that sense the model was derived from the zero-scattering condition within the dipole approximation, but does not invoke any additional long-wavelength approximation. As a result, it captures the effects of spatial dispersion and predicts a finite effective refractive index and anti-resonances in the electric permittivity and the magnetic permeability, in agreement with numerical finite-element calculations [37].

2.4 Equivalent-Circuit Theory and Transmission-Line Theory

To understand some underlying physical mechanisms of behaviors of MMs sometimes conceptual understandings are necessary. The novel properties of MMs can be analyzed theoretically by solving the Maxwell's equations. If we can solve the Maxwell's equations analytically, it is possible to understand the underlying physical mechanisms of behaviors of MMs thoroughly. Because the shape of the unit metallic patch of MMs can be very complicated; however, it is not an easy task to solve the Maxwell's equations analytically. Although we can solve the Maxwell's equations numerically and the results of numerical simulations can afford very valuable information, which can help us to understand the underlying physical mechanisms of behaviors of MMs, sometimes it is not easy to elucidate some systematic dependence of properties on, for instance, the geometrical parameters of unit cell, such as periodicity, length of metallic CW, distance between structures in a unit cell, thicknesses of dielectric substrate and metallic patterns, and so on. When analytic solution is not available, one of powerful tools to understand the systematics at least conceptually is to use the so-called equivalent-circuit theory, in which the metallic patterns are replaced by equivalent inductors and capacitors, and their inductances and capacitances can be found analytically by using the geometric parameters of metallic patterns. Through the equivalent inductances and capacitances the resonance frequencies can approximately obtained. By changing the geometric parameter of metallic patterns, systematic variations can be predicted and compared with the results of experiments. Although the equivalent-circuit theory is in general a conceptual tool, sometimes it can provide even quantitative results.

Zhou et al. [38] applied the equivalent-circuit theory to the sample, whose unit cell was composed of two CWs deposited on both sides of dielectric substrate, forming a CW pair (CWP), and four wires, two on each side of dielectric substrate, forming two CWPs. By knowing facts that the inductance of parallel plates is given by $L = \mu_0(l \cdot t_s)/w$, where l is the length of the CWs, w is the width, and t_s is the distance between the short wires, and the capacitance C of the CWPs is given by $C = \varepsilon_r\varepsilon_0(l \cdot w)/4t_s$, where ε_r is the permittivity of the region between CWs, the magnetic-resonance frequency f_m is given by

$$f_m = \frac{1}{2\pi\sqrt{LC}} = \frac{1}{\pi}\frac{1}{l\sqrt{\varepsilon_0\varepsilon_r\mu_0}} = \frac{c_0}{\pi l\sqrt{\varepsilon_r}}. \tag{2.55}$$

This equation implies that f_m is inversely proportional to the length of CWPs. The width or the separation between them do not play a role. This kind of knowledge cannot be obtained by simple numerical simulations, but it can provide valuable insights for design and structural optimization of MMs. The same strategy was employed by Lam et al. [39] to understand the dependence of the magnetic-resonance frequency on the width of CW in CWP medium.

Later, Lam et al. [40] applied the same method with a realistic elaboration. Because this is a good example how the equivalent-circuit theory works, a detailed discussion will be reproduced here. To find the effective capacitance (C_{eff}) and inductance (L_{eff}), they set up differential equations for the charge density and the current density, and solved the equations to get the total induced charge and current by integration. It was assumed that the electric field of incident EM wave was along the length of CW with length l, width w and thickness t_f, and the x, the y and the z axes were in the direction of the width, the length, and the thickness, respectively. Because the charge density is the function of y and the current density is given by

$$\mathbf{J} = J(y)\hat{\mathbf{y}} = \sigma E_0\hat{\mathbf{y}} - \alpha\nabla\rho = \sigma E_0\hat{\mathbf{y}} - \alpha\frac{\partial\rho}{\partial x}\hat{\mathbf{y}}, \tag{2.56}$$

where σ is the electrical conductivity of CW, $\hat{\mathbf{y}}$ is the unit vector along y axis, and α is a proportionality constant, which will be determined later. The first (second) term on the right-hand side is the current driven by electric force on the charge carriers (driven by the diffusion of the charge carriers due to density gradient). The continuity condition provides the relationship

$$\nabla\cdot\mathbf{J} + \frac{\partial\rho}{\partial t} = \frac{\partial J}{\partial y} + \frac{\partial\rho}{\partial t} = -\alpha\frac{d^2\rho}{dy^2} + \frac{\partial\rho}{\partial t} = 0. \tag{2.57}$$

Assuming that $\frac{\partial\rho}{\partial t} = -\omega\rho$. The differential equation for determining the charge density is

$$\frac{d^2\rho}{dy^2} + \frac{\omega}{\alpha}\rho = 0. \tag{2.58}$$

The time dependence of field and current can be ignored without loss of the generality hereafter. Since the charge density is antisymmetric about the center of CW, the charge density can be written as

$$\rho = \rho_o \sin\left(\sqrt{\frac{\omega}{\alpha}}y\right) \tag{2.59}$$

and the current density

$$\mathbf{J} = \sigma E_o\hat{\mathbf{y}} - \alpha\rho_o\sqrt{\frac{\omega}{\alpha}}\cos\left(\sqrt{\frac{\omega}{\alpha}}y\right)\hat{\mathbf{y}}. \tag{2.60}$$

Because the charge distribution and the current density are entirely confined inside CW, multiplying $\mathbf{r}$ by the continuity equation (2.57) and integrating over the volume of CW, the following expression can be obtained;

$$\int \mathbf{r}\left[\nabla \cdot \mathbf{J} + \frac{\partial \rho}{\partial t}\right] d^3\mathbf{r} = \int \mathbf{J} d^3\mathbf{r} + \frac{\partial \mathbf{p}}{\partial t} = 0, \tag{2.61}$$

where $\mathbf{p} \equiv \int \mathbf{r}\rho d^3\mathbf{r}$ is the dipole moment of the charge distribution. After some algebra, the charge and the current densities are given by

$$\rho = \rho_o \sin\left(\frac{\pi}{l} y\right), \tag{2.62}$$

and

$$\mathbf{J} = \frac{4\alpha\rho_o}{l}\left[1 - \frac{\pi}{4}\cos\left(\frac{\pi}{l} y\right)\right]\hat{\mathbf{y}}, \tag{2.63}$$

respectively. From these sources the electric and the magnetic fields produced by 21 × 21 CWPs can be calculated by directly applying the Coulomb's law and the Biot-Savart's law, respectively. Since these fields are periodic along the x and the y direction and the electric and the magnetic energies are symmetric about the origin, we calculate the fields only in the first quadrant confined by $a_x \times a_y$ and $0 \leq z \leq 25$ cm. The effective capacitance C_{eff} and inductance L_{eff} are obtained from the following relations;

$$U_{\text{E}} = \frac{1}{2}\int \varepsilon E^2 dv = \frac{Q^2}{2C_{\text{eff}}}$$

and

$$U_{\text{B}} = \frac{1}{2}\int \frac{B^2}{\mu} dv = \frac{1}{2} L_{\text{eff}} I^2.$$

Finally, the electric and the magnetic resonance frequencies come to be

$$\omega_{\text{e}} = \frac{1}{\sqrt{L_{\text{e,eff}} C_{\text{e,eff}}}}$$

and

$$\omega_{\text{m}} = \frac{1}{\sqrt{L_{\text{m,eff}} C_{\text{m,eff}}}},$$

respectively.

The calculational results are displayed in Fig. 2.7. The magnetic resonance frequency does not change appreciably with the distance between CWP layers (a_z), while the electric one changes significantly. As a_z approaches 4 mm, the electric resonance frequency keeps decreasing and reaches the magnetic resonance

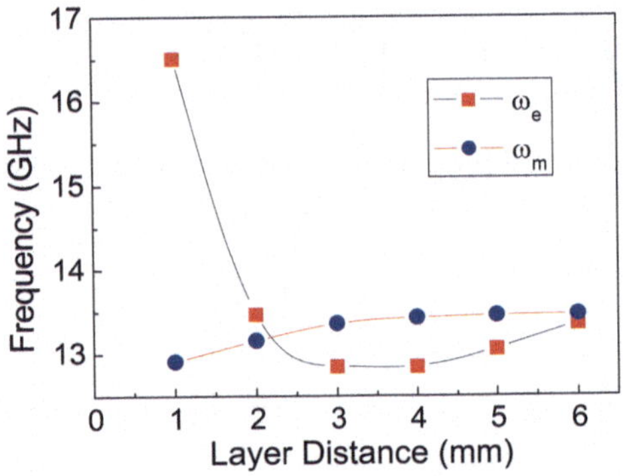

Fig. 2.7 Calculated electric and magnetic resonance frequencies as a function of the distance between CWP layers for 3-layer structure. The *solid lines* are guide to the eyes only. Reproduced from [40] with permission

frequency. Then the electric-resonance frequency increases again as a_z increases further. These calculational results successfully reproduce the experimental results.

The equivalent-circuit theory can also be applied to elucidate the influence of surface resistance on the frequency responses of MMPAs [41]. During the calculation the equivalent impedance of a lossy front surface could be represented as a series of *RLC* circuits. It was shown that the wide-band or narrow-band absorbers could be obtained by changing the surface resistance, and it was claimed that this method could be served as a guideline for the fast and efficient design of MM absorbers.

A very similar approach is TLT, which is based on the mapping of field components (i.e., **E** and **H**) in the medium to the complex voltages and currents of the equivalent *LC* circuits. The equivalent *LC* circuits are distributed one- or two-dimensionally. It is well known that ε and μ can be modeled using distributed *LC* circuits. TLT was originally developed by Heaviside [42].

When ordinary metallic, mostly Cu, wires carry a low-frequency alternating currents (ACs), it can be assumed that the voltage difference between two electrical terminals is negligible. For example, a 60-Hz EM wave has a wavelength of 5000 km, which is much longer than the size of ordinary electrical and electronic equipments. Even it is quite longer than most of the distances between a power plant to equipments using electricity generated by the plant. Therefore, it is reasonable to assume that the voltage difference between two electrical terminals is negligible. This is the basis of the long-wavelength approximation or the quasi-static circuit theory. In the long-wavelength approximation the dimension of electric equipment is negligibly small. If a certain amount of charge carriers are injected through a terminal it is nearly immediately ejected through the other tegminal. The charge density inside the transmission line does not change with time, which is the quasi-static condition. If the frequency of AC is increasing; however, we should give up such assumption at a certain frequency, where the length of wire carrying ACs is comparable to or larger than a significant fraction of the wavelength corresponding to the transmitting AC. As a consequence, it is not possible to use metallic wires to transmit currents at frequencies higher than radio frequency, because the energy is dissipated as radio waves, resulting in power losses, unless the metallic wires are carefully made to match the characteristic impedance of wires with the input impedance in a form of, for example, coaxial cables. Especially, any discontinuity in a current-carrying wire causes reflections, which hinder the propagation of signal through the wire. In the microwave range and higher, the energy loss in metallic wires as transmission lines is intolerable. In that case a waveguide can be used as a transmission line.

In TLT a two-port, or four-terminal network generally represents a transmission line, which connects a generator circuit at the sending-end port to load circuit at the receiving-end port. Each port consists of two terminals. Figure 2.8a shows a typical equivalent circuit used in TLT. Here, unlike the case of quasi-static circuit theory, the length of the transmission line is very important. The transmission line is further modelled as an infinite series of infinitesimal two-port elements, which represent an infinitesimally small segment of transmission line, as shown in Fig. 2.8b. The infinitesimal two-port element consists of a resistor-inductor segment, in which the resistor with resistance of dR and the inductor with inductance of dL are connected in series, and a conductor-capacitor element, in which the capacitor with capacitance of dC and the conductor with conductance of dG are connected in parallel. The potential difference across the infinitesimal element is

$$d\tilde{V}(x) = -(dR + i\omega dL)\tilde{I}(x) \tag{2.64}$$

and the infinitesimal current is

$$d\tilde{I}(x) = -(dG + i\omega dC)\tilde{V}(x). \tag{2.65}$$

Because the infinitesimal values of resistance, capacitance, conductance and inductance ares proportional to the infinitesimal length of the infinitesimally small segment of transmission line, i.e.,

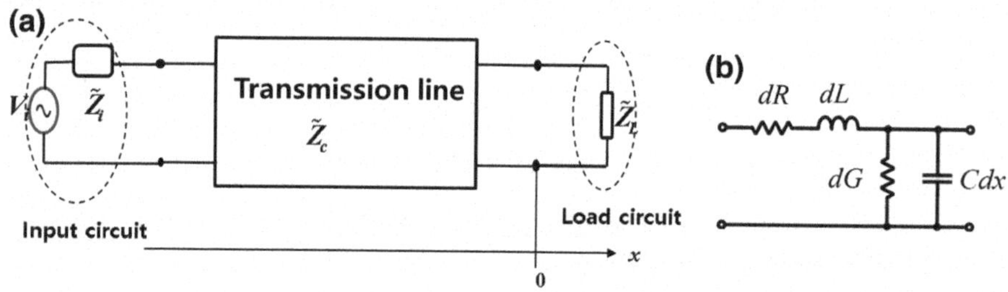

Fig. 2.8 Equivalent circuit used in the transmission-line theory

$$\begin{aligned} dR &= \bar{R}dx, \\ dL &= \bar{L}dx, \\ dC &= \bar{C}dx, \\ dG &= \bar{G}dx, \end{aligned} \tag{2.66}$$

where $\bar{R}$ is the distributed resistance in Ω/m, $\bar{L}$ is the distributed inductance in H/m, $\bar{C}$ is the capacitance between two conductors in F/m, and $\bar{G}$ is the conductance of dielectric material separating two conductors in S/m. Equations (2.64) and (2.65) can be rewritten as

$$\frac{d\tilde{V}(x)}{dx} = \tilde{\beta}^2\tilde{I}(x) \tag{2.67}$$

and

$$\frac{d\tilde{I}(x)}{dx} = \tilde{\beta}^2\tilde{V}(x), \tag{2.68}$$

respectively, where $\tilde{\beta} \equiv \sqrt{(\bar{R}+i\omega\bar{L})(\bar{G}+i\omega\bar{C})}$ is the propagation constant. The characteristic impedance, which is defined as the ratio of the amplitude of a single voltage wave to its current wave, is

$$\tilde{Z}_c = \sqrt{\frac{\bar{R}+i\omega\bar{L}}{\bar{G}+i\omega\bar{C}}}. \tag{2.69}$$

By solving these so-called telegrapher's equations the input impedance is given by

$$\tilde{Z}_i = \tilde{Z}_c\frac{\tilde{Z}_L + \tilde{Z}_c\tanh\left(\tilde{\beta}l\right)}{\tilde{Z}_c + \tilde{Z}_L\tanh\left(\tilde{\beta}l\right)}, \tag{2.70}$$

where l is the length of the transmission line. For lossless transmission, in which $\bar{R}=0$ and $\bar{G}=0$, the propagation constant is purely imaginary $\tilde{\beta} = ik$. The input impedance becomes

$$\tilde{Z}_i = \tilde{Z}_c\frac{\tilde{Z}_L + i\tilde{Z}_c\tan(kl)}{\tilde{Z}_c + i\tilde{Z}_L\tan(kl)} \tag{2.71}$$

If the length of transmission line is the quarter wavelength, the input impedance for lossless transmission is given by

$$\tilde{Z}_i = \frac{\tilde{Z}_c^2}{\tilde{Z}_L}, \tag{2.72}$$

because $k = 2\pi/\lambda$. For Salisbury screen the load impedance is 0 because the rear metal plate corresponds to the short circuit. Then, the lossy sheet is open.

2.5 Numerical Tcchniques

There are several techniques, such as plane-wave expansion and spherical-wave expansion, finite-difference time-domain (FDTD) method, transfer-matrix method (TMM), finite-element method (FEM), method of moment and so on, to solve the Maxwell's equations numerically. Many commercial softwares to calculate various properties of MMs, such as absorption, reflection and transmission spectra, field distributions, surface-current distributions, energy-density distributions, and so on, are also available. In this Section FDTD, FEM, and TMM will be discussed briefly. For thorough discussions of individual techniques, see Davidson [43].

2.5.1 *Finite-Difference Time-Domain Method*

FDTD method is the most popular method in MM simulations. In FDTD the numerical differentiation is based on the finite difference in both space and time. Starting point of FDTD is the Taylor expansion of any function as

$$f(x_0 + \Delta x) = f(x_0) + \left.\frac{df}{dx}\right|_{x=x_0} \Delta x + \frac{1}{2}\left.\frac{d^2 f}{dx^2}\right|_{x=x_0} (\Delta x)^2 + \frac{1}{3!}\left.\frac{d^3 f}{dx^3}\right|_{x=x_0} (\Delta x)^3 + \cdots \tag{2.73}$$

To apply FDTD the computational domain should be defined, first. The computational domain is a spatial domain where the simulation will be carried out. The computational domain is divided into small meshes, at which finite differences are calculated. From the Taylor expansion we can get the central-difference equation at the ith spatial mesh and at the nth time step as

$$f(x_i + \Delta x)|_{t_n} + f(x_i - \Delta x)|_{t_n} = 2f(x_i)|_{t_n} + (\Delta x)^2 \left.\frac{\partial^2 f}{\partial x^2}\right|_{x_i, t_n} + \frac{(\Delta x)^4}{12} \left.\frac{\partial^4 f}{\partial x^4}\right|_{x_i, t_n} + \cdots, \tag{2.74}$$

where i and n are integers, and Δx is the length of mesh. The second derivative is approximately given by

$$\left.\frac{\partial^2 f}{\partial x^2}\right|_{x_i,t_n} = \frac{f(x_i+\Delta x)|_{t_n} + f(x_i-\Delta x)|_{t_n} - 2f(x_i)|_{t_n}}{(\Delta x)^2} + \mathcal{O}\left[(\Delta)^2\right]. \tag{2.75}$$

By using definitions of $f_i^n \equiv f(x_i)|_{t_n}$, $f_{i+1}^n \equiv f(x_i+\Delta x)|_{t_n}$ and $f_{i-1}^n \equiv f(x_i-\Delta x)|_{t_n}$ the equation above becomes

$$\left.\frac{\partial^2 f}{\partial x^2}\right|_{x_i,t_n} = \frac{f_i^{n+1} + f_i^{n-1} - 2f_i^n}{(\Delta x)^2} + \mathcal{O}\left[(\Delta x)^2\right]. \tag{2.76}$$

This notation will be used for the wave equation

$$\frac{\partial^2 u}{\partial x^2} = \frac{1}{c^2}\frac{\partial^2 u}{\partial t^2}, \tag{2.77}$$

where u can either the electric field E or the magnetic field H, as

$$\frac{u_{i+1}^n + u_{i-1}^n - 2u_i^n}{(\Delta x)^2} + \mathcal{O}\left[(\Delta x)^2\right] = \frac{1}{c^2}\left\{\frac{u_i^{n+1} + u_i^{n-1} - 2u_i^n}{(\Delta t)^2} + \mathcal{O}\left[(\Delta t)^2\right]\right\}. \tag{2.78}$$

The updated finite difference is

$$u_i^{n+1} \cong \frac{u_{i+1}^n + u_{i-1}^n - 2u_i^n}{(\Delta x)^2}(c\Delta t)^2 - u_i^{n-1} + 2u_i^n. \tag{2.79}$$

To solve the Maxwell's equation through FDTD, the fields are calculated in the so-called Yee cell (see Fig. 2.9). Because the electric field should satisfy the relationship

$$\frac{\partial \mathbf{E}}{\partial t} = \frac{1}{\varepsilon}[\nabla \times \mathbf{H} - (\mathbf{J} + \sigma\mathbf{E})], \tag{2.80}$$

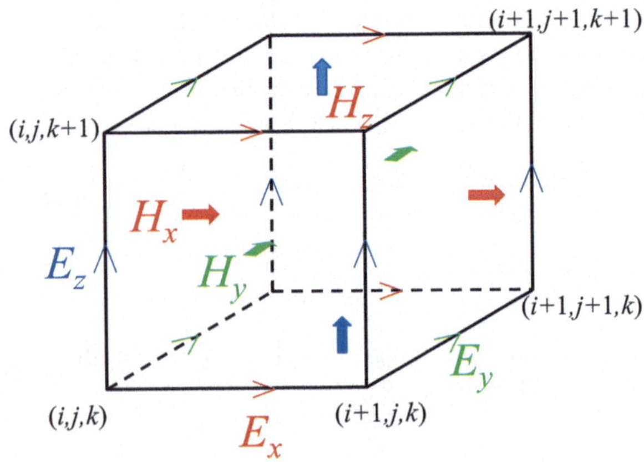

Fig. 2.9 Standard Cartesian Yee cell used for FDTD

where **J** is, if any, the current source and σ is the electrical conductivity, the x-component of electric field in the Yee cell is given by

$$\frac{\partial E_x}{\partial t} = \frac{1}{\varepsilon}\left(\frac{\partial H_z}{\partial y} - \frac{\partial H_y}{\partial z}\right) - (J_x + \sigma E_x) \tag{2.81}$$

and thus the central-difference equation becomes

$$\begin{aligned}
&\frac{E^{n+1/2}_{x,\{i,j+1/2,k+1/2\}} - E^{n-1/2}_{x,\{i,j+1/2,k+1/2\}}}{\Delta t}\\
&= \frac{1}{\varepsilon_{i,j+1/2,k+1/2}}\left(\frac{H^n_{z,\{i,j+1,k+1/2\}} - H^n_{z,\{i,j,k+1/2\}}}{\Delta y} - \frac{H^n_{y,\{i,j+1/2,k+1\}} - H^n_{y,\{i,j+1/2,k\}}}{\Delta z}\right.\\
&\qquad\left. - J^n_{x,\{i,j+1/2,k+1/2\}} - \sigma_{i,j+1/2,k+1/2}E^n_{x,\{i,j+1/2,k+1/2\}}\right),
\end{aligned} \tag{2.82}$$

where j and k are integers. Here, we utilize the relationship

$$\left.\frac{\partial f}{\partial t}\right|_{i\Delta x, j\Delta y, k\Delta z, n\Delta t} = \frac{f^{n+1/2}_{i,j,k} - f^{n-1/2}_{i,j,k}}{\Delta t} + \mathcal{O}\left[(\Delta t)^2\right] \tag{2.83}$$

and the so-called leapfrog method for time stepping. The other components can easily be found by appropriately modifying (2.82). The magnetic field can be obtained similarly as

$$\begin{aligned}
&\frac{H^{n+1/2}_{x,\{i,j+1/2,k+1/2\}} - H^{n-1/2}_{x,\{i,j+1/2,k+1/2\}}}{\Delta t}\\
&= \frac{1}{\mu_{i,j+1/2,k+1/2}}\left(\frac{E^n_{z,\{i,j+1,k+1/2\}} - E^n_{z,\{i,j,k+1/2\}}}{\Delta y} - \frac{E^n_{y,\{i,j+1/2,k+1\}} - E^n_{y,\{i,j+1/2,k\}}}{\Delta z}\right.\\
&\qquad\left. - M^n_{x,\{i,j+1/2,k+1/2\}} - \bar{\sigma}_{i,j+1/2,k+1/2}H^n_{x,\{i,j+1/2,k+1/2\}}\right),
\end{aligned} \tag{2.84}$$

where **M** is, if any, the magnetization source and $\bar{\sigma}$ is the magnetic conductivity. The choice of fineness of time step (Δt) and meshes (Δx, Δy and Δz) is very important. If they are chosen to be too fine it is not only time-consuming but also may be numerically instable. The numerical instability is probably due to the truncation of digits, which is inevitable in electronic computer. If they are chosen to be too crude it might be fast, but the results might not be acceptable. By a rule of thumb the so-called Courant-Friedrichs-Lewy condition [44]

$$\Delta t = \frac{\Delta x}{\sqrt{D}c}, \tag{2.85}$$

where D is the spatial dimension, should be followed. Although the FDTD method is very versatile and spectrum of broad frequency range can be obtained at once, it requires a lot of computing resources. In addition, ε and μ cannot be uniquely determined at the interface.

2.5.2 *Finite-Element Method*

FEM is another numerical technique to solve the boundary-value problems of partial differential equations (PDEs). To carry out FEM the spatial region of interest should be divided into properly-chosen meshes or grids. The shape of mesh can be triangle in 2-D space; however, it can be any form. In 3-D space it can be tetrahedral, hexahedral, prism, or pyramidal shape. All meshes should be connected to adjacent meshes. Then, those connected meshes produce nodes, which are usually points at which 3 or more meshes meet. Meshes should be chosen in such a way that there should be no empty space between meshes; however, the total outer surface of all outermost meshes should not necessarily identical to the real boundaries.

Next step is to choose basis functions which can be used to set up the approximate solution at each node. The basis function can be piecewise linear in a mesh. In these days, the basis functions are usually piecewise polynomial. At each node the coefficients of polynomial are set to satisfy the BCs. The field quantities are interpolated in the mesh. The number of equations are finite. If the size of meshes is infinitesimally small these approximate solution can be exact, however the degrees of freedom would be infinite, which is impossible to attack. If the size of mesh is macroscopically large, but small enough to make approximate solution reliable, the degrees of freedom become finite. It is the origin of the name, 'finite'-element method.

There are several differences between FEM and finite-difference method (FDM), such as FDTD. FEM approximates the solution, while FDM approximates the PDEs. FEM can handle very complex shape of boundary of structure, while FDM can handle relatively simple shape only. Most of all, FEM is easy to implement. There are; however, several disadvantages. Because FEM approximates solution itself with a piecewise polynomial at each node, it is not possible to find a general closed-form solution, from which the dependence of the investigated system's response on the variation of parameters is studied systematically. FEM inherently contains errors, therefore, it is fatal if users make mistakes.

All numerical methods inherently contain errors because of the so-called truncation error of computer. Because computer can handle only a finite magnitude of number itself and the number of digits is also limited, it is not possible to handle numbers, which are beyond the largest magnitude the computer can handle. Furthermore, if the number of digits is infinite (irrational numbers) or larger than the allowed number of digits, it is inevitable to truncate the rest of digits. If the truncations are accumulated during the process the results can be fatally inaccurate. Besides this truncation error, FEM contains a few sources of inherent errors. They deal simplified geometry, polynomial interpolation of field quantities, simple integration technique, such as Gaussian quadrature, and so on. There are also common mistakes made by users, such as wrong choice of type of elements, distorted elements, insufficient support which prevents, for instance, all rigid-body motion, inconsistent units, etc. In spite of these drawbacks, FEM is one of the most popular numerical methods in numerous fields of engineering because of easiness of implementation and versatility in handling very complex surfaces.

2.5.3 Transfer-Matrix Method

TMM is in general a technique for finding solutions in statistical mechanics. If the partition function $\mathscr{Z}$ of a system under consideration can be written as

$$\mathscr{Z} = \mathbf{u}_0 \cdot \left(\prod_{i=1}^{N} T_i \right) \cdot \mathbf{u}_{N+1}, \tag{2.86}$$

where $\mathbf{u}_0$ and $\mathbf{u}_{N+1}$ are vectors of dimension D and $\mathbf{T}_i$'s are $D \times D$ matrices, which are called transfer matrices. In some cases, the partition function is simply

$$\mathscr{Z} = \mathrm{Tr}\left(\prod_{i=1}^{N} \mathbf{T}_i \right), \tag{2.87}$$

where 'Tr' stands for trace of a matrix.

This basic idea of transfer matrix can be applied in the field of optics. Especially, TMM is very simple and accurate method when the propagation of EM wave through a stratified medium. For the simplicity an example of 1-D case will be given. Extension to higher dimensions is straightforward. Consider a system schematically shown in Fig. 2.10. If there are N layers of different media, then there are $N + 1$ interfaces including frontal and rear surfaces facing vacuum or air. The EM wave travels along z-axis. We can further assume that the EM wave is linearly polarized along the x-axis without loss of generality. In the jth layer the electric field can be represent as

$$E_j(z) = \overrightarrow{E}_j e^{ik_j z} + \overleftarrow{E}_j e^{-ik_j z}, \tag{2.88}$$

where k_j is the wave vector of EM wave in the jth layer, satisfying $k_j = 2\pi/\lambda_0 n_j$ with the refractive index of the jth layer, n_j, and the vacuum wavelength of incident

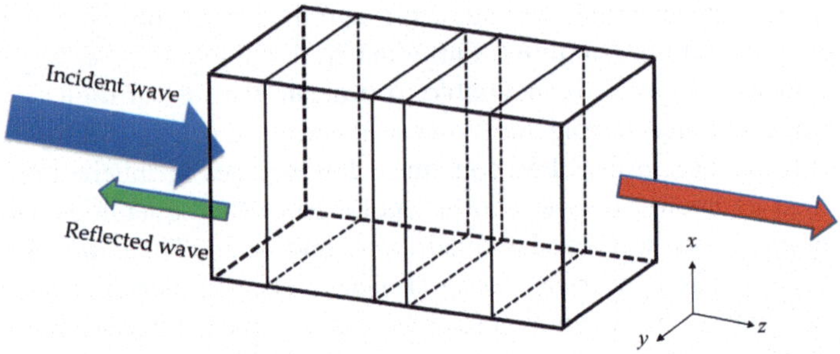

Fig. 2.10 Schematic of multilayered structure upon which an EM wave is incident, reflected and transmitted

EM wave, λ_0. $\overleftarrow{E}_j$ ($\overrightarrow{E}_j$) is the amplitude of wave traveling along the direction of increasing z (decreasing z). This field and its derivative with respect to z,

$$\xi_j(z) \equiv \frac{\partial E_j(z)}{\partial z} = ik_j \overrightarrow{E}_j e^{ik_j z} - ik_j \overleftarrow{E}_j e^{-ik_j z}, \tag{2.89}$$

should be continuous at interfaces. The continuity of the electric field and its derivative at a certain interface requires

$$\begin{pmatrix} E_j\big|_{z=z_{j+1}} \\ \xi_j\big|_{z=z_{j+1}} \end{pmatrix} = \begin{pmatrix} E_{j+1}\big|_{z=z_{j+1}} \\ \xi_{j+1}\big|_{z=z_{j+1}} \end{pmatrix}, \tag{2.90}$$

where z_j is location of the interface between $(j-1)$th and jth layers. Because

$$E_j\big|_{z=z_j} = \overrightarrow{E}_j e^{ik_j z_j} + \overleftarrow{E}_j e^{-ik_j z_j} \tag{2.91}$$

and

$$\xi_j\big|_{z=z_j} = ik_j(\overrightarrow{E}_j e^{ik_j z_j} - \overleftarrow{E}_j e^{-ik_j z_j}), \tag{2.92}$$

we can get

$$\overrightarrow{E}_j = \frac{ik_j E_j\big|_{z=z_j} + \xi_j\big|_{z=z_j}}{2ik_j} e^{ik_j z_j} \tag{2.93}$$

and

$$\overleftarrow{E}_j = \frac{ik_j E_j\big|_{z=z_j} - \xi_j\big|_{z=z_j}}{2ik_j} e^{-ik_j z_j}. \tag{2.94}$$

By substituting these into

$$E_j\big|_{z=z_{j+1}} = \overrightarrow{E}_j e^{-ik_j(z_j+d_j)} + \overrightarrow{E}_j e^{ik_j(z_j+d_j)} \tag{2.95}$$

and

$$\xi_j\big|_{z=z_{j+1}} = ik_j(\overleftarrow{E}_j e^{-ik_j(z_j+d_j)}, -\overrightarrow{E}_j e^{ik_j(z_j+d_j)}) \tag{2.96}$$

where $d_j \equiv z_{j+1} - z_j$ is the thickness of jth layer, we can get the following relationship;

$$\begin{pmatrix} E_{j+1}\big|_{z=z_{j+1}} \\ \xi_{j+1}\big|_{z=z_{j+1}} \end{pmatrix} = \mathbf{M}_j \begin{pmatrix} E_j\big|_{z=z_j} \\ \xi_j\big|_{z=z_j} \end{pmatrix} \tag{2.97}$$

with

$$\mathbf{M}_j = \begin{pmatrix} \cos(k_j d_j) & \frac{1}{k_j}\sin(k_j d_j) \\ k_j \sin(k_j d_j) & \cos(k_j d_j). \end{pmatrix}. \tag{2.98}$$

Therefore,

$$\begin{pmatrix} E_{N+1}\big|_{z=z_N} \\ \xi_{N+1}\big|_{z=z_N} \end{pmatrix} = \mathbf{M}_N \ldots \mathbf{M}_j \ldots \mathbf{M}_2 \mathbf{M}_1 \begin{pmatrix} E_0\big|_{z=z_0} \\ \xi_0\big|_{z=z_0} \end{pmatrix} = \mathbf{M} \begin{pmatrix} E_0\big|_{z=z_0} \\ \xi_0\big|_{z=z_0} \end{pmatrix}, \tag{2.99}$$

where $\mathbf{M} \equiv \prod_{j=1}^{N} \mathbf{M}_j$ is the transfer matrix.

Assuming that $z<0$ in the region from which the incident wave comes,

$$E_0(z) = \bar{E}_0 e^{ikz} + \tilde{r}\bar{E}_0 e^{-ikz} \quad (\text{if } z<0) \tag{2.100}$$

and

$$E_{N+1}(z) = \tilde{t}\bar{E}_0 e^{ikz} \quad (\text{if } z > d), \tag{2.101}$$

where $\tilde{r}$ and $\tilde{t}$ are the reflection and the transmission, respectively, coefficients of the whole system, which is immersed in a vacuum or air and $d = d_1 + d_2 + \cdots + d_N$ is the total thickness of multilayered system. $\tilde{r}$ and $\tilde{t}$ can be found as

$$\tilde{r} = \frac{(M_{21} + k^2 M_{12}) + ik(M_{22} - M_{11})}{(-M_{21} + k^2 M_{12}) + ik(M_{22} + M_{11})} \tag{2.102}$$

and

$$\tilde{t} = 2ike^{-ikd} \frac{M_{11}M_{22} - M_{12}M_{21}}{-M_{21} + k^2M_{12} + ik(M_{11} + M_{22})}, \quad (2.103)$$

where M_{ij}'s are the $\{ij\}$-component of the transfer matrix.

TMM is inherently unstable because of the existence of backward propagating waves. If the wave vector is complex a decay of EM wave occurs as it propagates. In TMM all waves are treated as forward propagating. Therefore, the backward wave is treated as a forward propagating wave which is exponentially growing. This leads to a numerical instability. To cure this problem we must calculate the Poynting vector associated with the modes and carefully inspecting the sign of the z component.

References

1. J. Jackson, *Classical Electrodynamics*, 3rd edn. (Wiley, 1998)
2. C. Brau, *Modern Problems in Classical Electrodynamics* (OUP USA, 2004)
3. J. Reitz, F. Milford, R. Christy, *Foundations of Electromagnetic Theory* (Pearson/Addison-Wesley, New York, 2009)
4. V.G. Veselago, Sov. Phys. Usp. **10**, 509 (1968)
5. J. Pendry, A. Holden, D. Robbins, W. Stewart, IEEE T. Microw. Theory **47**, 2075 (1999)
6. D.R. Smith, W.J. Padilla, D.C. Vier, S.C. Nemat-Nasser, S. Schultz, Phys. Rev. Lett. **84**, 4184 (2000)
7. G.T. Ruck, D.E. Barrick, W.D. Stuart, A.C.K. Krichbaum, *Radar Cross Section Handbook*, vol. 2 (Plenum, 1970)
8. K. Ford, B. Chambers, I.E.E.E.T. Antenn, Propag. **56**, 133 (2008)
9. D. Holtby, K. Ford, B. Chambers, IET Radar Sonar Nav. **5**, 483 (2011)
10. G. Manara, A. Monorchio, R. Mittra, Electron. Lett. **35**, 1400 (1999)
11. DJM electronics. URL http://www.djmelectronics.com/rf-absorber.html
12. O. Kazmina, V. Suslyaev, M. Dushkina, B. Semukhin, IOP Conf. Ser.: Mater. Sci. Eng. **81**, 012036 (2015)
13. G. Li, T. Xie, S. Yang, J. Jin, J. Jiang, J. Phys. Chem. C **116**, 9196 (2012)
14. S.E. Lee, O. Choi, H.T. Hahn, J. Appl. Phy. **104**, 033705 (2008)
15. X. Ji, M. Lu, F. Ye, Q. Zhou, Adv. Mech. Eng. Appl. **3**, 294 (2012)
16. W.W. Salisbury. Absorbent body for electromagnetic waves, US Patent 2599944 (1952)
17. B.A. Munk, *Frequency Selective Surfaces: Theory and Design* (Wiley-Interscience, 2000)
18. H. Severin, I.R.E.T. Antenn, Prop. **4**, 385 (1956)
19. W. Dällenbach, W. Kleinsteuber, Hochfreq. u Elektroak **51**, 152 (1938)
20. B. Munk, P. Munk, J. Pryor, IEEE T. Antenn. Prop. **55**, 186 (2007)
21. H.Y. Zheng, X.R. Jin, J.W. Park, Y.H. Lu, J.Y. Rhee, W.H. Jang, H. Cheong, Y.P. Lee, Opt. Express **20**, 24002 (2012)
22. N.I. Landy, S. Sajuyigbe, J.J. Mock, D.R. Smith, W.J. Padilla, Phys. Rev. Lett. **100**, 207402 (2008)
23. C. Hu, X. Li, Q. Feng, X. Chen, X. Luo, Opt. Express **18**, 6598 (2010)
24. L. Li, J. Wang, H. Du, J. Wang, S. Qu, Z. Xu, AIP Adv. **5**, 017147 (2015)
25. J.W. Lee, M.A. Seo, J.Y. Sohn, Y.H. Ahn, D.S. Kim, S.C. Jeoung, C. Lienau, Q.H. Park, Opt. Express **13**, 10681 (2005)

26. P. Tuong, J. Park, V. Lam, W. Jang, S. Nikitov, Y. Lee, Opt. Comm. **295**, 17 (2013)
27. A. Vora, J. Gwamuri, N. Pala, A. Kulkarni, J.M. Pearce, D.. Güney, Sci. Rep. **4** (2014)
28. M.A. Ordal, L.L. Long, R.J. Bell, S.E. Bell, R.R. Bell, R.W. Alexander, C.A. Ward, Appl. Opt. **22**, 1099 (1983)
29. J.Y. Rhee, Y.J. Yoo, K.W. Kim, Y.J. Kim, Y.P. Lee, J. Electromagn. Waves Appl. **28**, 1541 (2014)
30. D.A.G. Bruggeman, Ann. Phys. (Leipzig) **24**, 636 (1935)
31. R. Landauer, AIP Conf. Proc. **40**, 2 (1978)
32. J. Garnett, Phil. Trans. R. Soc. Lond. **203**, 636 (1904)
33. W.S. Weiglhofer, A. Lakhtakia, B. Michel, Microw. Opt. Techn. Lett. **15**, 263 (1997)
34. J.Y. Rhee, X. Wang, B.N. Harmon, D.W. Lynch, Phys. Rev. B **51**, 17390 (1995)
35. S. Resink, E. Hondebrink, W. Steenbergen, Opt. Express **22**, 3564 (2014)
36. Y. Wu, J. Li, Z.Q. Zhang, C.T. Chan, Phys. Rev. B **74**, 085111 (2006)
37. B.A. Slovick, Z.G. Yu, S. Krishnamurthy, Phys. Rev. B **89**, 155118 (2014)
38. J. Zhou, L. Zhang, G. Tuttle, T. Koschny, C.M. Soukoulis, Phys. Rev. B **73**, 041101 (2006)
39. V.D. Lam, J.B. Kim, S.J. Lee, Y.P. Lee, J.Y. Rhee, Opt. Express **15**, 16651 (2007)
40. V.D. Lam, J.B. Kim, N.T. Tung, S.J. Lee, Y.P. Lee, J.Y. Rhee, Opt. Express **16**, 5934 (2008)
41. Y.Q. Pang, Y.J. Zhou, J. Wang, J. Appl. Phys. **110** (2011)
42. O. Heaviside, *Electrical Papers*, vol. 2 (Macmillan, 1894)
43. D.B. Davidson, *Computational Electromagnetics for RF and Microwave Engineering*, 2nd edn. (Cambridge University Press, Cambridge, New York, 2010)
44. R. Courant, K.O. Friedrichs, H. Lewy, Math. Ann. **100**, 32–74 (1928)

Chapter 3
MMPAs Operating in Different Frequency Ranges

Abstract In this chapter, metamaterial perfect absorbers (MMPAs) are introduced by categories according to their operating frequencies: MMPA operating in the GHz, the THz, and the infrared (IR)/visible ranges. Although the existence of famous THz gap, below which the response of device is dominantly electronic, while above which it is dominantly photonic, the basic principle of operation of MMPA in the whole frequency range is nearly identical to each other, except for small details, for example, localized surface-plasmon polaritons which come into play in the IR/visible range because their energies correspond to IR/visible. Therefore, a simple scale down of the structure operating at GHz is possible for operating in higher-frequency ranges. For individual frequency ranges considered in this Chapter the broadband behavior, polarization independence, tunability and omni-directional behavior are discussed individually. In the near future MMPA will play a key role in the field of photonics and optoelectronics.

3.1 MMPAs Operating in the GHz Range

In the first perfect-absorption (PA) metamaterial (MM) by Landy et al. [1], they theoretically and experimentally realized a perfect MM absorber in the GHz range. The suggested structure was composed of conducting electric ring resonator on the front side and cut wire on the rear one, separated by a dielectric substrate (as shown in the left panel of Fig. 3.1). By manipulating the magnetic-resonance coupling without changing the electric ring resonator, they were able to decouple ε and μ and individually tune the resonance. Then, perfect impedance matching was achieved in the MM medium whose high dielectric loss in the insulator was used as the heat generating factor. They shows absorption peak of 96 % at 11.48 GHz and of 88 % at 11.5 GHz for simulation and experiment, respectively (see right panel of Fig. 3.1). Their results theoretically revealed that the absorption mechanism could be understood by two reasons; the Ohmic surface loss and the dielectric loss outside the conductor. The latter loss occurs in between two MM elements where the

Y.P. Lee et al., *Metamaterials for Perfect Absorption*,
Springer Series in Materials Science 236, DOI 10.1007/978-981-10-0105-5_3

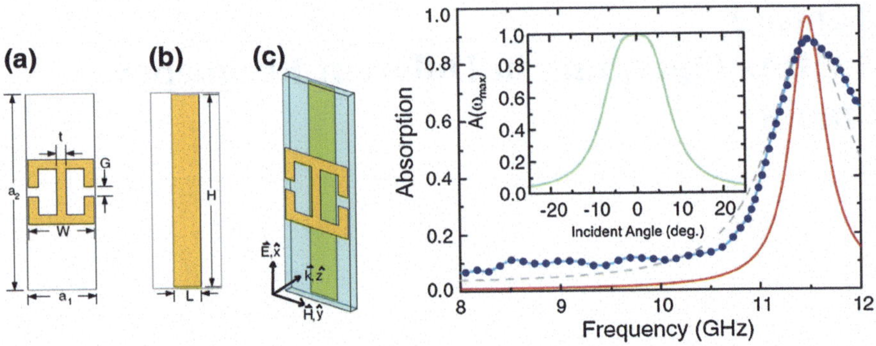

Fig. 3.1 (*Left*) **a** Electric resonator and **b** cut wire. **c** Unit cell with axes indicating the propagation direction. (*Right*) *Main panel* shows the simulated and measured $A(\omega)$ lines. The *gray dashed* $A(\omega)$ line is a Gaussian-weighted average which approximates the assembly error. *Inset* shows the simulated angular dependence of the $A(\omega)$ at $\omega_{\max}$. Reproduced from [1] with permission

electric field is large. They suggested that the method to obtain PA is to fabricate two metallic layers: the metallic pattern on the front layer is for the minimized reflection of electromagnetic (EM) wave by impedance matching with the incident medium, and the metallic layer on the back side is for blocking the transmission, usually made of continuous metallic layer. However, the general PA MMs are prepared for a specific frequency. Evidently, the PA MMs with a single-band high absorption are applicable in a limited area. Therefore, the research on multi/broadband, polarization-independent, incident-angle-independent and frequency-tunable MM perfect absorbers (MMPAs) is necessary.

In 2010, Li et al. realized a dual-band metamaterial absorber in the microwave region [2]. The structure consisted of electric ring resonator on the front layer and cut wire on the rear one, separated by a dielectric FR-4 substrate (see the top-left panel of Fig. 3.2). The simulated and the experimental results showed that the absorber had two PA points at 11.15 and 16.01 GHz (right panel of Fig. 3.2). Absorption at different polarizations of incident EM wave were measured to be over 97 % for low-frequency peak and 99 % for high-frequency one. They verified that the PA was induced by the impedance ($\tilde{Z}$) matching: absorption peaks at 11.15 and 16.01 GHz. The real relative impedance is near unity, $\mathrm{Re}\left[\tilde{Z}\right] \sim 1$, which means $\tilde{Z} \approx Z_0$ (Z_0 is the impedance of the free space) and the imaginary part is minimized, $\mathrm{Im}\left[\tilde{Z}\right] \sim 0$, such that the reflection is nearly zero (shown in the bottom-left panel of Fig. 3.2).

The triple-band MMPA at GHz was theoretically and experimentally investigated by Shen et al. [3]. As shown in the left panel of Fig. 3.3, the unit cell of triple-band absorber also consisted of three layers. The top layer was an array of three nested copper closed-ring resonators, which was primarily responsible for the electric response to the incident EM wave. The bottom layer was a copper plane, which is used to zero the transmission and is responsible for the magnetic response, exciting antiparallel surface current to that of the top layer. The middle layer was

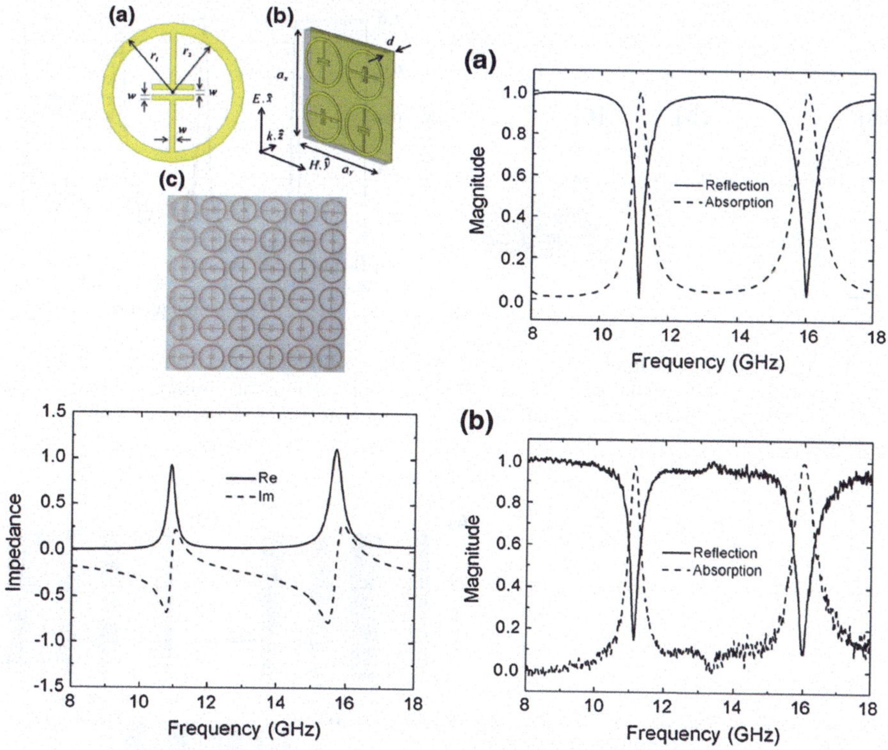

Fig. 3.2 (*Top left*) **a** Front view of the single electric split-ring resonator. **b** Perspective view of the unit cell. **c** Photograph of a portion of the experimentally-realized absorber (*top layer*). The single electric split-ring resonator next to the other is rotated by 90°. (*Right*) **a** Simulated and **b** measured reflection and absorption as a function of frequency under normal incidence with electric field polarized along x axis. (*Bottom left*) Real part and imaginary part of relative impedance extracted from the simulation. [2] Reproduced courtesy of the Electromagnetics Academy

FR-4 dielectric substrate. The simulated absorption spectrum had three peaks at frequencies of 4.02, 6.75 and 9.24 GHz with absorption of 99, 96 and 98 %, respectively. The full width at half-maximum (FWHM) of each absorption peak is 0.16, 0.23 and 0.33 GHz, respectively. These results indicated that the FWHM is wider for the absorption peak at higher frequency. The agreement between experiment and simulation was good (see the right panel of Fig. 3.3).

Park et al. [4] theoretically and experimentally reported the polarization-independent quadruple-band MMPAs in the GHz range by using donut structures. The unit-cell structure consists of four-kind-patterned metallic layer at the front and metallic plane at the back, separated by a dielectric FR-4 substrate (see the top-left panel of Fig. 3.4). As shown in the bottom panel of Fig. 3.4, the experimental absorption spectra are good agreement with the simulation ones. The experimental results revealed quad peaks at 6.5, 7.4, 9.2 and 11.0 GHz with absorption of 97, 97,

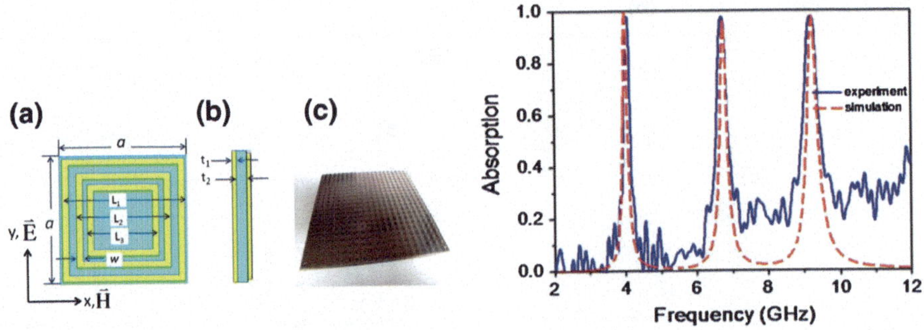

Fig. 3.3 (*Left*) Schematic structure of the microwave absorber. **a** Front and **b** side view of the unit cell. **c** Photograph of the triple-band absorber. (*Right*) Comparison of the absorption spectra of the triple-band MM absorber between experimental and simulated results. Reproduced from [3] with permission

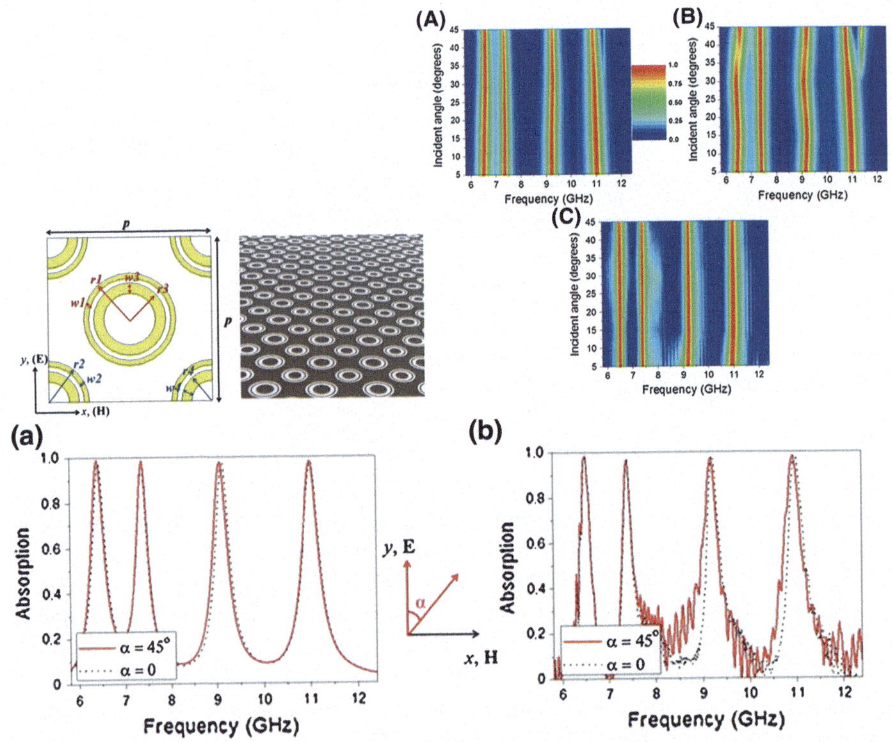

Fig. 3.4 (*Top left*) Schematics of the proposed structure of MMPA for quad peaks (*left*) and photo of the fabricated sample (*right*). (*Bottom*) **a** Simulated and **b** measured absorption spectra of quad-peak MMPA according to polarization. (*Top right*) Dependence of the simulated absorption spectra on incident angle for **A** TE mode and **B** TM mode. **C** Corresponding measured absorption spectra for TE mode. Reproduced from [4] with permission

98 and 98 %, respectively. The origin of absorption mechanism turned out to be the magnetic resonance, which was caused by the antiparallel surface currents between front and back plates. Moreover, the suggested MM structure could be no difference between TE and TM polarizations (shown in the top-right panel of Fig. 3.4). Furthermore, they explained that the absorption was nearly independent of the incident angle up to 45°. This result can be clarified by the fact that the absorption-peak position of magnetic resonance is not much changed according to the angle of incidence.

On the other hand, Zheng et al. [5] reported the dual-band perfect absorber by both simulation and experiment, which could tune the resonance frequency. They suggested a tunable dual-band perfect absorber by combining extraordinary optical-transmission effect and Fabry-Pérot cavity resonance. These ideas and the mechanism of the absorption are completely different from the absorber based on the near-field interaction. The dual-band-absorber structure was composed of two separated metallic layers, which were deposited on dielectric FR-4 substrates (see Fig. 3.5a). Figure 3.5b presents the absorption spectra of the structure with

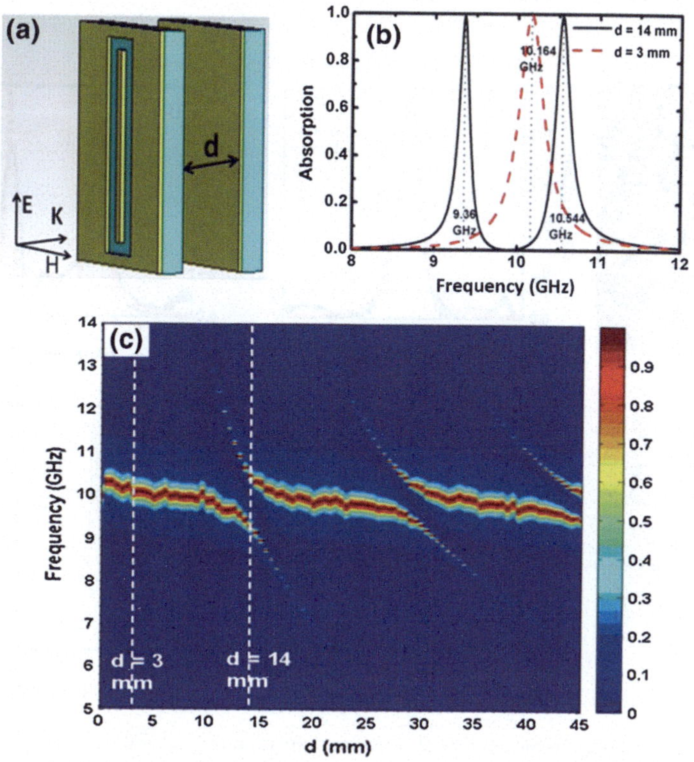

Fig. 3.5 **a** Unit cell of dual-band absorber, **b** simulated absorption spectra of the dual-band absorber with $d = 14$ mm and single-band absorber with $d = 3$ mm and **c** absorption as a function of d and frequency. Reproduced from [5] with permission

$d = 3$ and 14 mm. For $d = 3$ mm only one absorption peak exists, while in the case of $d = 14$ mm there are two peaks of nearly PA at 9.36 and 10.54 GHz. Figure 3.5c shows the absorption spectra according to d. With decreasing the distance from $d = 14$ to $d = 3$ mm, dual-band absorption turns into the single-band absorption. In addition, the resonance frequency of absorption can be controlled by varying distance *d*. The experimental results were coincident with those of the simulation (Fig. 3.6b, c of the top panel). In order to extend the range of applications, they also designed a polarization-insensitive dual-band absorber by disk-type structure (shown in Fig. 3.6a of the bottom panel). Figure 3.6b of the bottom panel presents three basic structures (structure A, B and C), relevant to structure D. Figure 3.6c of the bottom panel reveals the absorption spectra by varying polarization angle. These results indicated that polarization-insensitive tunable dual-band perfect absorber was realized because of the symmetry of structure.

Tuong et al. also reported polarization-insensitive and polarization-controlled dual-band absorption in the GHz range [6]. The unit cell of the MM absorbers was composed of three layers, metallic dual rings at the front and metallic plane at the back, separated by a dielectric FR-4 layer (shown in Fig. 3.7a of the left panel). They suggested two types of structures: one of them is the separated dual-ring

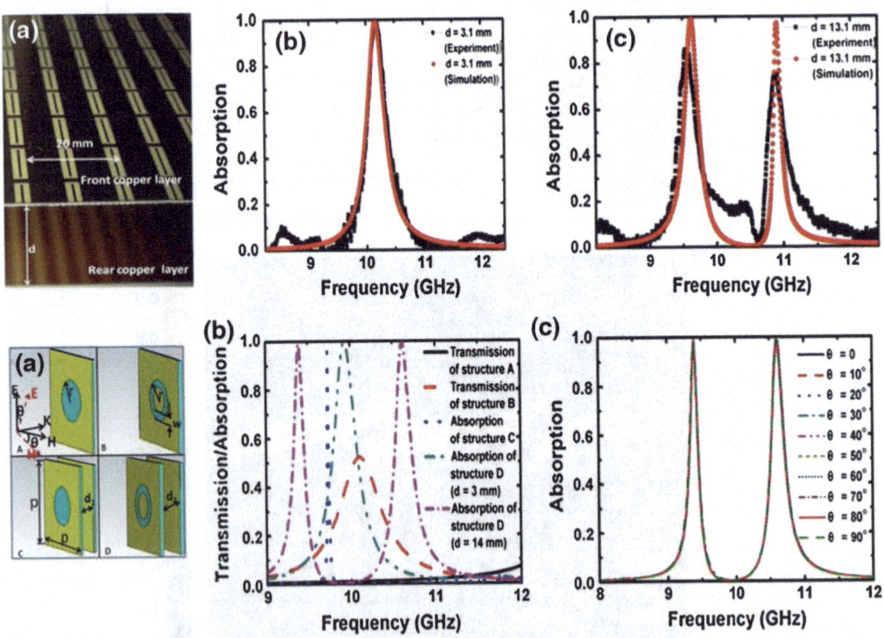

Fig. 3.6 (*Top*) **a** Photo of the fabricated structure. Experimental and simulated absorption spectra with **b** $d = 3.1$ and **c** 13.1 mm. (*Bottom*) **a** Unit cells of structure A, B, C and D. Related parameters are set to be d = 14 mm, w = 0.5 mm, r = 3.6 mm and periodicity as 22 mm. **b** Transmission spectra of various structures. **c** Absorption spectra of structure D according to polarization angle. Reproduced from [5] with permission

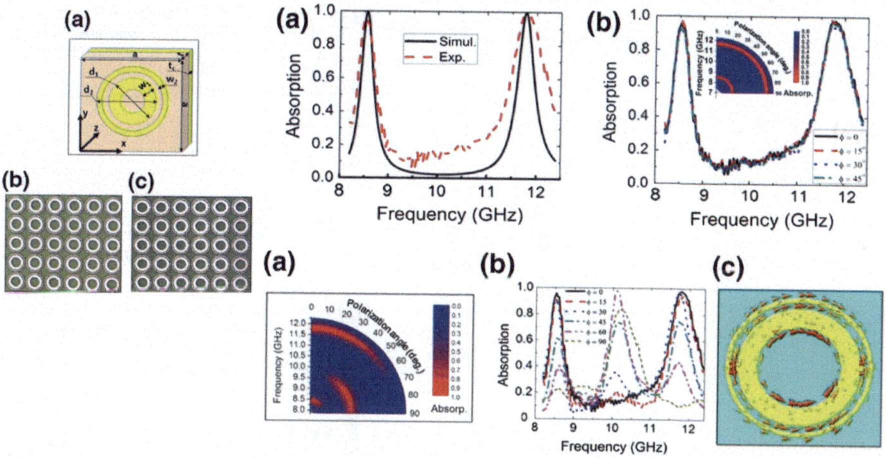

Fig. 3.7 (*Left*) **a** Simulated design of the unit cell of dual-band absorption MM. Photos of **b** the dual-ring MM and **c** the connected dual-ring MM. (*Top right*) **a** Simulated and experimental absorption spectra. **b** Absorption spectra according to polarization angle (simulated results in the *Inset*). (*Bottom right*) **a** Simulated and **b** experimental results of polarization-controllable absorption at GHz frequencies. **c** Induced surface currents at the combined resonance mode. Reproduced from [6] with permission

structure and the other is the connected dual-ring structure (see Fig. 3.7b, c of the left panel, respectively). For the separated dual-ring structure, two nearly perfect absorption peaks are evident. Two nearly PA peaks corresponded to the resonances at lower and higher frequencies, induced by the coupling of outer and inner rings with metallic plane at the back, respectively (Fig. 3.7a of the top-right panel). Nearly perfect dual-band absorption was obtained at 8.55 and 11.78 GHz. The perfect dual-band absorption exhibits polarization-insensitive properties by utilizing the symmetrical geometry of the MM, which are plotted in Fig. 3.7b of the top-right panel. For the connected dual-ring structure, on the other hand, the dual-band absorber can be optimized to be sensitive according to the polarization of incident EM wave. Figure 3.7a, b of the bottom-right panel show the simulation and the experiment of the polarization-controllable absorption in the GHz region, respectively. It is seen that the dual-band absorption is gradually reduced by the polarization angle. This is switched off at an angle of 90°. Moreover, the combined-mode resonance appears at 10 GHz. The combined-mode resonance is induced by the surface currents at connection between the inner and the outer rings as in Fig. 3.7c of the bottom-right panel.

In general, multi-band MM absorber should need the corresponding kinds of meta-patterns. On the other hand, Yoo et al. [7] suggested theoretically and experimentally polarization-independent dual-band perfect absorber utilizing only one kind of metallic pattern. Figure 3.8a, b of the top-left panel show schematics for the two suggested MM absorber structures, disk and donut types, respectively. The experimental spectra are similar to the simulation (top-right panel of Fig. 3.8).

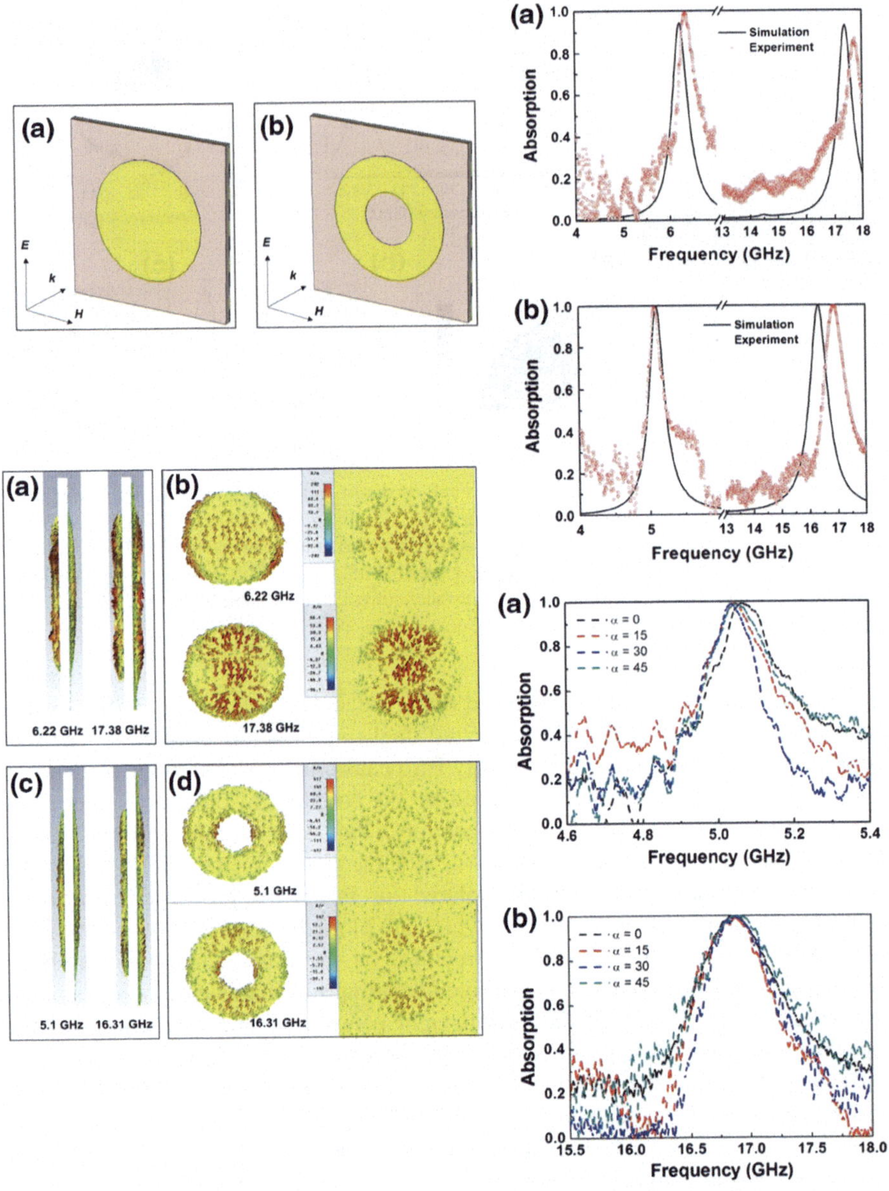

Fig. 3.8 (*Top left*) Unit cells for the MM absorbers of **a** disk-type and **b** donut-type structures. (*Top Right*) Simulated and measured absorption spectra for **a** disk-type and **b** donut-type structures. (*Bottom left*) Measured absorption spectra of donut-type absorber for various polarizations. (*Bottom right*) Induced surface currents for (**a** and **b**) the disk-type and (**c** and **d**) the donut-type absorbers. Reproduced from [7] with permission

The absorption of the low-frequency peak for donut-type structure shows absorption of 99.1 % at 5.06 GHz and of 99.6 % at 5.10 GHz for experiment and simulation, respectively. On the other hand, the experimental high frequency peak shows the absorption of 99.9 % at 16.88 GHz. They explained the absorption mechanism of two absorption peaks by using the surface currents at each absorption peaks (shown in Fig. 3.8 of the bottom-left panel). The origin of lower-frequency absorption peak was caused by the fundamental magnetic resonance. On the other hand, the higher-frequency absorption peak was induced by the third harmonic magnetic resonance. In addition, the multi-plasmonic perfect absorptions for donut-type are polarization-independent (bottom-right panel of Fig. 3.8). The designed MMPA could provide absorption higher than 99 % at the third-harmonic magnetic resonance as well as the fundamental one, independently from polarization.

In order to realize a broad-band MM absorber, many researcher have tried to obtain the broadband absorption of MM utilizing various meta-structures. In 2012, Ding et al. reported ultra-broadband MM absorber in the GHz range [8]. The unit cell of the MM absorber was periodic array of quadrangular frustum pyramids with a homogeneous metal film as the ground to block the transmission (shown in Fig. 3.9). These pyramids possess various resonance absorption modes at multi-frequencies, of which the overlapping leads to the total absorption of the incident wave over the ultra-wide spectral band. As shown in Fig. 3.9a of the right panel, the simulated absorption spectra reveals the ultra-broadband absorption in the 8-14 GHz. The experimental absorption is shown in Fig. 3.9b of the right panel.

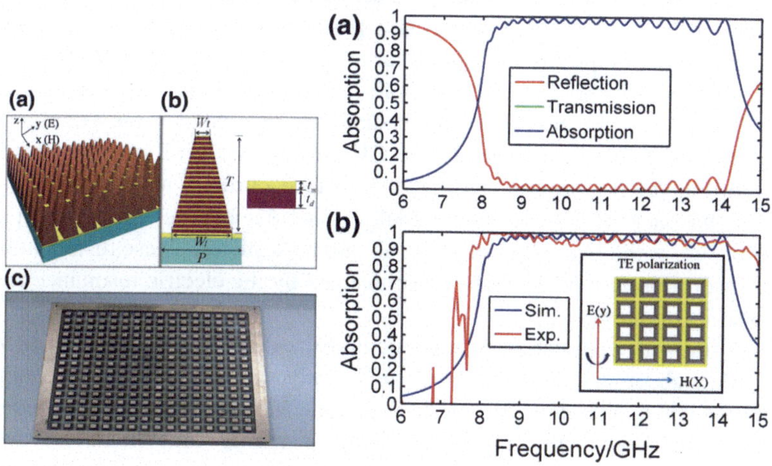

Fig. 3.9 (*Left*) Design and fabrication of a microwave ultra-broadband MM absorber. **a** Three-dimensional illustration of the simulated MM absorber, **b** schematic of the MM-absorber unit cell and **c** photograph of the fabricated sample. (*Right*) **a** Numerical simulation: absorption, reflection, and transmission. **b** Comparison between the simulated absorption and experimental absorption. *Inset* in (**b**) shows the configuration of the incident wave. Reproduced from [8] with permission

The agreement between experimental and simulated results is very good, considering the tolerance in the fabrication and assembly. On the other hand, Kim et al. suggested the combination between the ultra-broadband and the third-harmonic magnetic resonance [9]. They also realized the polarization-independent and dual-broadband MM absorbers at microwave frequencies. The periodic meta-atom array consisted of metal-dielectric-multilayer truncated cones (see the top-left panel of Fig. 3.10). In simulation, the absorption was over 90 % in 3.93–6.05 GHz, and 11.64–14.55 GHz (in the middle-left panel of Fig. 3.10). They demonstrated not only one broadband absorption from the fundamental magnetic resonances, but additional broadband absorption in high-frequency range using the third-harmonic resonance (see the top-right panel of Fig. 3.10 and the bottom-right panel of Fig. 3.10). The corresponding experimental absorption bands over 90 % were 3886.08, 99,510.46 and 118,613.84 GHz, respectively. As shown in the bottom-left panel of Fig. 3.10, the suggested multilayer truncated-cone structure also revealed polarization-angle independence due to the multilayered circular structures.

On the other hand, Dung et al. proposed broadband MM absorber based on a new attempt: the meta patterns were connected by embedded resistor [10]. As shown in the top-left panel of Fig. 3.11, the meta-atom consisted of three layers: metallic squares connected by two resistors (Fig. 3.11b of the top-left panel) at the front and a metallic plane at the back, separated by a dielectric FR-4 layer. First, the properties of the MM absorber without connecting the chip resistor (see the bottom-left panel of Fig. 3.11). Two absorption peaks are evident; the first absorption peak (94 % at 7.2 GHz) is induced by the fundamental magnetic resonance and the second absorption peak (98 % at 18.7 GHz) is caused by the third magnetic resonance. Two unit cells were connected with a chip resistor, as in the right panel of Fig. 3.11a. Interestingly, new absorption peak appears additionally at 13 GHz, and the total power absorption turns out to be higher than 90 % in a band of 1.3 GHz from 12.3 to 13.6 GHz in both the simulated and experimental results (Fig. 3.11b of the right panel). To understand the mechanism, they demonstrated the distribution of surface current at 13.6 GHz in Fig. 3.11c of the right panel. The surface-current distribution shows that the surface currents at the corners of the front and the back plates are antiparallel. At the same time, the strong parallel currents occurs at the two edges of the front and back plates. These results indicated that the absorption around 13 GHz was enhanced by the electric resonance and can be broader in frequency.

In addition, many researchers have studied the deep-sub-wavelength MMs. In general MM absorbers, the unit-cell size is only 1/3–1/5 of the wavelength of incident EM wave. Yoo et al. reported the flexible MM absorber for low frequency, based on small-size unit cell [11]. Using a planar and flexible MM, the low-frequency perfect absorber was realized with very small unit-cell size in snake-shape structure. As schematically depicted in the top-left panel of Fig. 3.12, the structure was composed of three layers, patterned copper layer, dielectric flexible Teflon layer and continuous copper layer. They chose the Teflon substrate for the GHz-range MM absorber, since the Teflon substrate is thin, flexible and elastic, as shown in Fig. 3.12c of the top-left panel. Figure 3.12a of the right panel

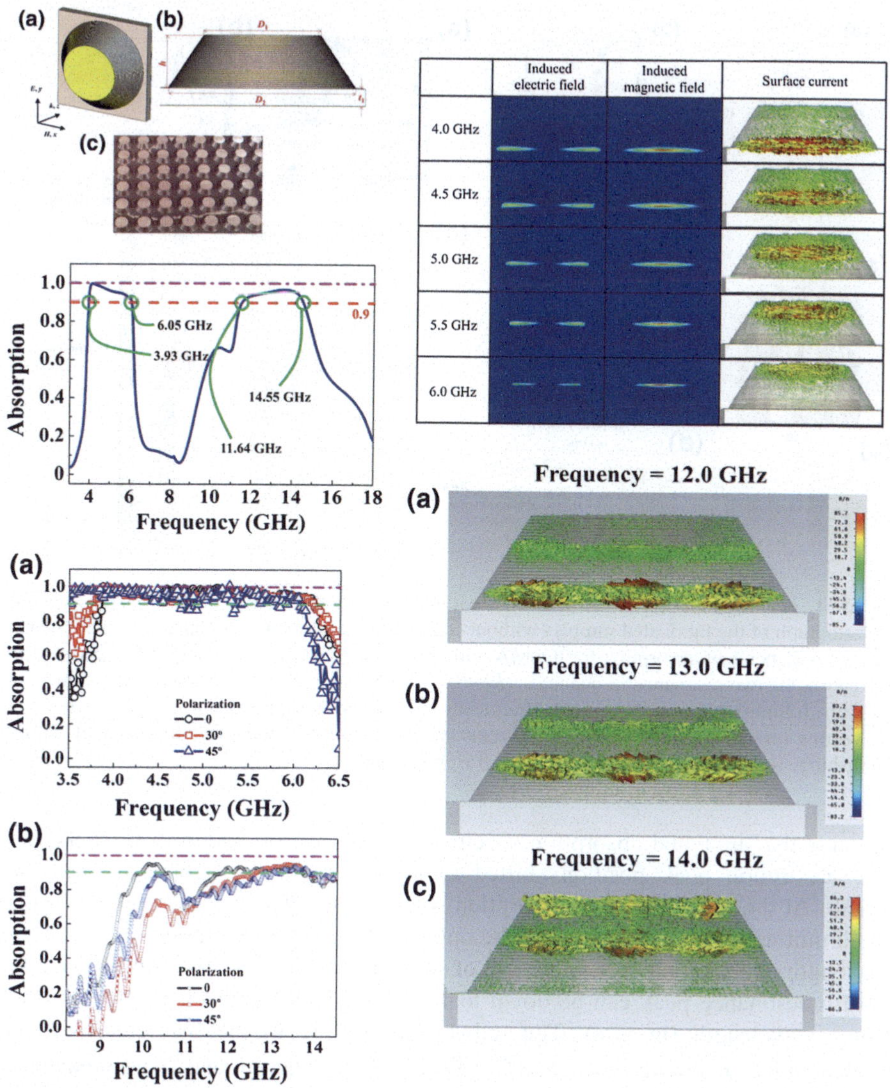

Fig. 3.10 (*Top left*) **a** Perspective view and **b** side view of the designed meta-atom. D_1 and D_2 are the top and the bottom diameters of circular structures. t_1 and h are the thickness of FR-4 and the height of truncated cone structure. **c** Photo of the fabricated sample. (*Middle left*) Simulated absorption spectrum of the truncated cone-structure absorber. (*Top right*) Simulated electric- and magnetic-field distributions at the central cross section of unit cell at 4.0, 4.5, 5.0, 5.5 and 6.0 GHz. *Left* and *center* columns are for the electric and the magnetic fields, respectively, and the *right one* shows the induced surface current. (*Bottom right*) Simulated surface current in the unit cell at specific frequencies: **a** 12.0, **b** 13.0 and **c** 14.0 GHz. (*Bottom left*) Measured absorption spectra according to polarization angle in **a** 3.5–6.5 and **b** 9.0–14.5 GHz. Reproduced from [9] with permission

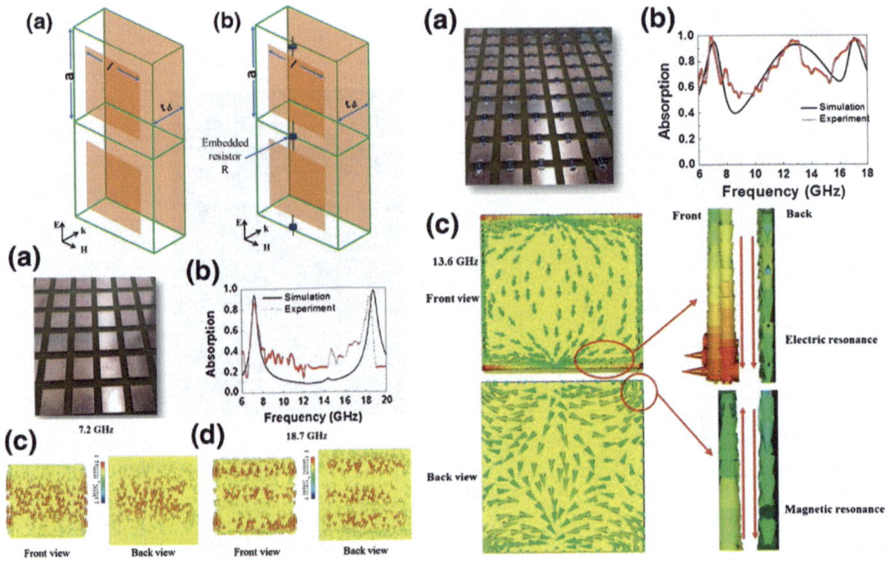

Fig. 3.11 (*Top left*) Design of the unit cell **a** without and **b** with embedded resistors. (*Bottom left*) **a** Photograph of the fabricated samples without connecting resistors. **b** Simulated and experimental frequency-dependent absorption for the MA without connecting resistors. **c** and **d** Induced surface currents at the two resonance frequencies. (*Right*) **a** Photo of the fabricated sample with connecting resistors. **b** Simulated and experimental frequency-dependent absorption for the MM absorber with connecting resistors. **c** Induced surface currents around the electric resonance frequency. [10] IOP Publishing. Reproduced with permission. All rights reserved

presents the simulated absorption spectra according to the length m of snake bar. The absorption peak position shifted to higher frequency with decreasing the length. At the same time, the absorption was reduced. These results indicated that small unit-cell size of absorber can be optimized for high absorption by increasing the length of snake bar. For the length of snake bar of 7.5 mm, the unit-cell size for the PA resonance peak can be down to 1/12 (at 2 GHz) with respect to the resonance wavelength (in mm). The red-shift according to the bar length can be explained by *LC*-circuit resonance $f = 1/(2\pi\sqrt{LC})$, where L and C are inductance and capacitance, respectively. The length of snake bar could be considered as the inductance L. The capacitance C is regarded as the interval between neighboring snake bars. When the length of snake bar increases, the frequency of resonance peak was shifted to lower frequency. As shown in Fig. 3.12c of the right panel, the experimental spectra were in good agreement with the simulation. Furthermore, they introduced MM absorber for the MHz range. As in the inset of Fig. 3.12 of the bottom left panel, the unit cell contains 5 snake bars. The resonance peak revealed nearly PA of 99.8 % at 403.3 MHz. For this case, the size of unit cell is 30-times smaller than the resonance wavelength (in mm).

Khuyen et al. reported two type of MMPA in MHz region [12]. The proposed MMPA was a periodic arrangement of the unit cell depicted in Fig. 3.13a of the

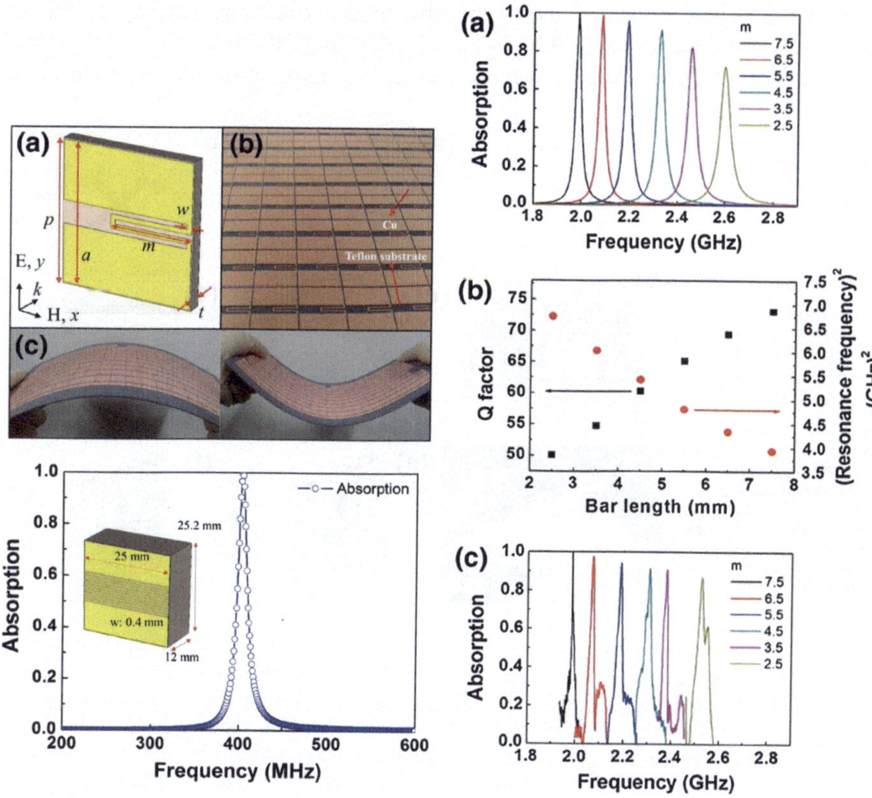

Fig. 3.12 (*Top left*) **a** Unit cell for the MM absorber in snake structure and **b** photo of the fabricated sample. **c** Photo to show the flexibility of the MM absorber for the GHz range. (*Right*) **a** Simulated absorption spectra, **b** *Q*-factor and square of resonance frequency and **c** measured absorption spectra, according to bar length *m*. (*Bottom left*) Simulated absorption spectra and the unit cell for the MMPA at 400 MHz. Reproduced from [11] with permission

top-left panel. A patterned metallic front layer was composed of four quarters of disk localized at corners, which were linked together by slender wires. The front layer was separated from a continuous back metallic layer by a dielectric spacer. Figure 3.13b, c of the left panel present the real samples of single-peak and dual-band MMPAs, respectively, which were fabricated by the conventional printed circuit-board process. Figure 3.13a of the right panel reveals the absorption spectrum of single-peak MMPA. The simulated and the experimental results are quite coincident. The nearly PA was obtained to be 98.6 and 99.7 % around 400 MHz in simulation and measurement, respectively. As shown in Fig. 3.13b, c of the right panel, the induced surface currents flow in bottom-up direction on the front layer. The induced charges are confined to the edges of the quarters of disk, and the corresponding mirror image of opposite charges is formed in the ground metallic plane. This result indicates that the origin of PA is the magnetic resonance.

Figure 3.14a of the left panel shows the absorption spectra by varying g_2 for the normally-incident plane wave. When $g_1 = g_2 = 0.15$ mm, the single absorption peak is observed at 405 MHz. When g_2 is increased, the single absorption is changed to

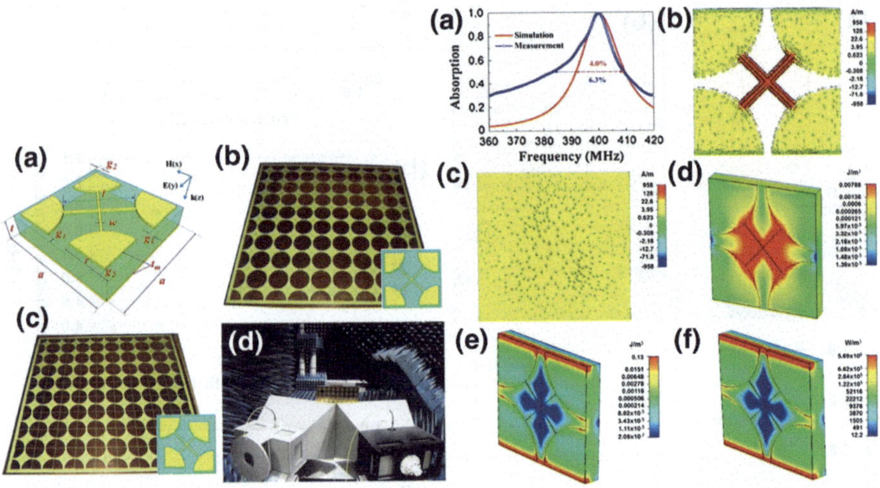

Fig. 3.13 (*Left*) **a** Unit cell of proposed MMPA. Photos of fabricated **b** single-peak and **c** dual-band sample. *Insets* in (**b**) and (**c**) are the unit cells, corresponding to single-peak and dual-band MMPA structures, respectively. **d** Experimental set-up for the measurement. (*Right*) **a** Simulated and measured absorption spectra of single-peak MMPA. Induced surface currents on **b** front and **c** back layers at 400 MHz. 3-dimensional distributions of **d** magnetic energy, **e** electric energy and **f** power loss at the resonant frequency. Reproduced from [12] with permission

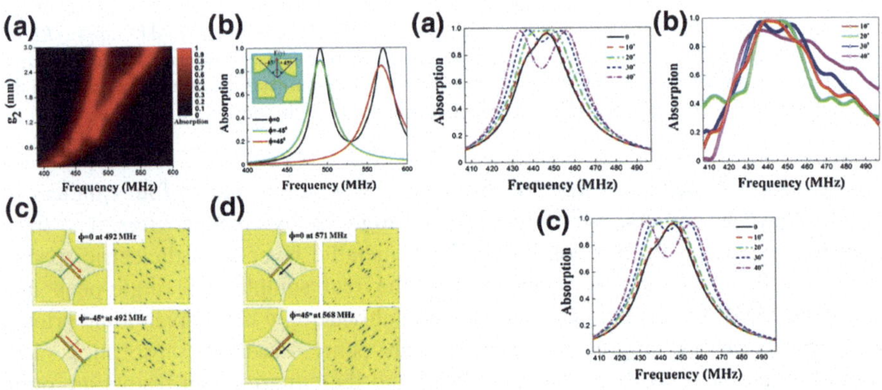

Fig. 3.14 (*Left*) **a** Dependence of the absorption on the gap (g_2) of proposed dual-band MMPA. **b** Simulated absorption spectra according to the polarization angle (ϕ) of EM wave. The *Inset* is the schematic view of varying the polarization angle. Induced surface currents on the front and the back metallic layers at two absorption peaks for **c** $\phi = -45°$ and **d** $\phi = 45°$. (*Right*) **a** Simulated and **b** measured absorption spectra of enhanced-bandwidth MMPA according to the incident angle of EM wave for TE polarization. **c** Simulated absorption spectra for TM polarization. Reproduced from [12] with permission

be the dual absorption peaks. When g_2 is changed to be 3.0 mm, two absorption peaks at 492 and 571 MHz are clearly observed due to the high asymmetry of structure. The absorption is 99.97 and 99.60 % at 492 and 571 MHz, respectively. Both absorption mechanisms are anti-parallel surface currents on the front and the back metallic layers as in Fig. 3.14c, d of the left panel. Figure 3.14a, b of the right panel present the simulated and measured absorption spectra according to the incident angle for TE polarization. The measured absorption spectra are in good agreement with the simulated ones. The absorption bandwidth is extended from 10 to 24 MHz at absorption of 90 % by increasing the incident angle from 10° to 30°. The absorption spectra according to the incident angle for TM polarization are similar to those of TE polarization (Fig. 3.14c of the right panel).

The EM wave in the GHz range, is used primarily for cell phone and military radar bands. In order to be applied to these application areas, therefore, MMPAs operating in the GHz range should overcome the problems relevant to the size of the unit cell, the flexibility, the thermal stability, etc. The deep-subwavelength (the lattice parameter ~$\lambda/10$) MMPAs should be investigated for immediate and versatile applications, since the lattice parameter increases for the MMPAs operating in the low-frequency range. Recently, the flexible MMPAs for GHz are also necessary, utilizing ink-jet printing and 3-dimensional printing. Moreover, the thermal stability of MMPAs are studied by employing various dielectric materials. Broadband MMPAs utilizing naturally-common materials such as tap water and so on, are being investigated. This might be more practical and cost-effective for the applications.

3.2 MMPAs Operating in the THz Range

International Telecommunication Union defines the terahertz (THz) radiation as the EM wave whose frequency ranges from 0.3 to 3 THz (1 THz = 10^{12} Hz, equivalent to the wavelength of 300 μm). In this Section, however, we redefine the THz radiation as the EM wave in the regime extending from 100 GHz to 10 THz as shown in Fig. 3.15. The latter is defined by the so-called 'THz gap'. THz radiation lies between the electronic response-dominating regime (microwave) and the photonic response-dominating (IR) regime. It implies that in this regime neither the

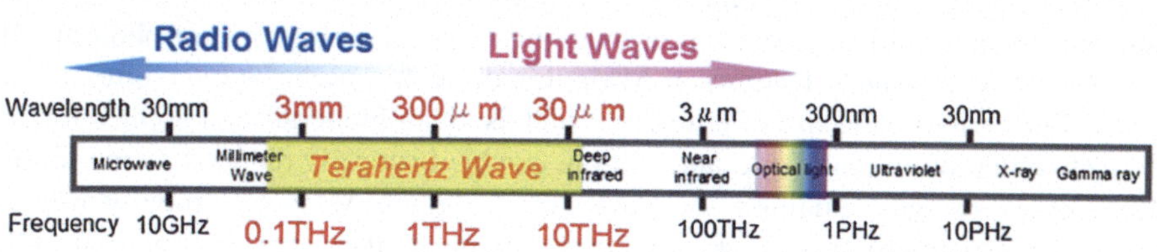

Fig. 3.15 THz regime of the EM wave lies between electronic (microwave) and photonic (IT) frequency, and extends from 100 GHz to 10 THz (Origin: http://www.fujitsu.com/global/about/resources/news/press-releases/2011/0912-01.html)

electronic nor the photonic responses predominates over the other. At frequencies of a few hundred GHz and lower, the motion of free electrons forms the basis of most EM devices where the classical EM theory is sufficient to describe them, especially, the circuit theory. On the other hand, at IR through optical/UV wavelengths, photon-based devices are dominant where the classical EM theory, especially, the Maxwell's equations, is still working, and sometimes the quantum-mechanical treatment of EM waves, i.e., photon, is inevitable. In between these two regions, there exists the so-called "THz gap," where the efficiency of electronic and photonic responses diminishes if frequency changes from the individual ends. Therefore, there is no efficient technique to generate and detect the THz radiation for practical use, such as telecommunication. In telecommunication THz radiation particularly reveals limited usefulness because some water-vapor absorption bands falls on the THz range, hindering to use it in long-distance communication. Hence, the THz regime is inarguably the least developed and the least understood portion of the whole EM spectrum both scientifically and technologically [13, 14]. Moreover, the lack of reliable device technology is one of the major constraints for the full exploitation of THz technology. In spite of these facts the THz frequency range of EM wave attracts great interest recently due to the potential applications such as biomedical diagnostics and homeland security where the advantages arising from the non-ionizing nature of THz ray and the capability to penetrate some specific materials like paper, clothes, etc., can be fully exploited [13]. Moreover, the THz radiation allows to detect substance-specific spectral features with a sub-mm diffraction-limited resolution, therefore becoming very attractive for quality and process control in non-destructive way and for environmental-monitoring implementations [15]. It is also known that THz radiation is very useful in short-distance communications, especially, in-door applications, due to its imaging capability and ability to be applied in the construction of high-bandwidth wireless networking systems. The response of natural materials, arising from interaction of the EM field with the electrons, forms the basis for the construction of most modern devices. However, the nature of the EM response of materials changes as a function of frequency.

MMs are prefect candidate for the practical application in the THz range because, by utilizing MMs, novel and impossible-in-nature properties of a system can be realized. In fact, the lack of efficient techniques to generate and detect the THz radiation does not mean the lack of strong interaction of the THz radiation with matters; however, there is no efficient way to improve the efficiency of emitters and detectors because of the absence of dominating responses, electronic or photonic. If we utilize the resonance features of MMs this difficulty might be overcome rather easily. Furthermore, the size of emitters, which is major hinderance of application of the THz radiation in real-world application, can be drastically reduced owing to the deep-subwavelength condition for MMs. Even though MMs are initially implemented in the microwave-frequency range owing to the large size of unit cell, leading to the simplicity in fabrication, MMs can be realized over a wide spectral range extending from the THz- to the visible-frequency range by scaling down the structural dimensions and by utilizing standard micro/nano fabrication technologies.

By appropriately 'diluting' the effective electron density plasmons can be excited, which are essential to the operations of MMs, in the targeted frequency range. The design flexibility associated with MMs provides a promising approach, from device perspective towards filling the THz gap [16].

It has been demonstrated that the MMs have potential importance in the THz regime, and various THz-MM devices have been reported, including PAs, THz modulators for both amplitude and phase, wave guides, electromagnetically-induced-transparency devices, broadband THz generators and detectors, and antennas. On some of these devices, more sophisticated characteristics such as tunability, multi-band or broadband spectral response, and incident-angle- and/or polarization-independent responses have been demonstrated. In this Section, recent progresses on THz MMPAs will be reviewed.

Various types of MM structures have been theoretically designed and experimentally demonstrated during the last couple of decades, for example, thin metallic wires [17, 18], Swiss rolls [19, 20], pairs of rods and crosses [21–23], fishnet structures [24–26], split-ring resonators (SRRs) [27–29], combination of U-shaped resonators [30], chiral structures [31] and other multilayered structures [32–34]. Among these resonator structures, SRRs are the canonical structures used in the majority of THz MMs to date [35–40].

Yen et al. [41] were the researchers, who introduced THz MM for the first time. They experimentally demonstrated THz MM by using a planar layer of double SRR array. In order to avoid the electric resonance attributed by the capacitance of the facing metallic stripes, as shown in Fig. 3.16, the electric field was aligned parallel

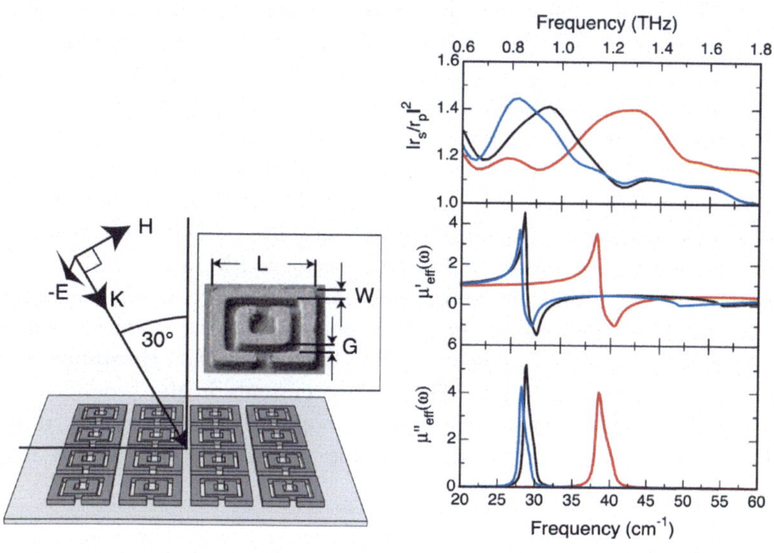

Fig. 3.16 (*Left*) Magnetically-coupled SRRs with different geometries excited by a 30° off-normal incident with an ellipsometer. (*Right*) Ratio of the magnetic to electric response (*top*), simulation results for the real (*middle*) and imaginary (*bottom*) magnetic-permeability functions. Reproduced from [41] with permission

to the two SRR gaps, and a strong magnetic response around 1 THz was attained. The operation of MMs based on the simple resonator structure usually results in a narrow spectral response due to the nature of the resonator. The resonance is defined as the oscillation of a certain physical quantity of a system, which is under an influence of external stimulus varying periodically with a certain frequency, showing a large amplitude at the frequency of external stimulus. If the so-called *Q*-factor of the system is large the amplitude of oscillation decreases rapidly if the frequency of external frequency is slight different from that of the resonance. The *Q*-factor depends on the loss or dissipation of input energy into the system, i.e., the higher the loss the lower the *Q*-factor. To achieve the perfect absorption the frontal surface of MMPA should have impedance matching with that of the free space, implying that the top layer of MMPA is lossless, as a consequence, revealing a very narrow resonance band. The magnetic response at THz frequencies can be achieved in a planar structure composed of nonmagnetic conductive-resonant elements because the magnetic response of ordinary material is usually very weak.

Recently, various efforts have been made on broadening the bandwidth of THz MMs by packing two or more resonators with different geometries into a single unit cell or by adapting dual-layer structure for multi-resonant responses. The above-mentioned methods of fabricating multi-band or broadband MMs have been widely used [42]. Wang et al. suggested a square-loop-circle SRR THz filter, which exhibits a dual stop bands [43]. There are two transmission dips close to zero before and after the central transmission peak located around 0.6 THz. They are relevant to a high absorption. Figure 3.17 shows an electric-SRR structures consisting of dual-band resonators and metallic grounded plane separated by a dielectric spacer layer, which exhibits dual-band absorption spectra [44]. Since these structures contain a dielectric spacer between resonant component and grounded metallic plane, the effect of spacer layer should be carefully considered in determining the spectral response of such MMPA structures to realize a controllable EM response of MMs by adjusting the geometries of electric and magnetic resonators individually to create a highly selective absorber over a narrow band at THz frequencies [45]. It should be particularly noted here that the name of the structure fabricated on the first layer, ELC resonator,[1] is misleading. Because the structure on the first layer is strongly coupled to the electric field of the incident EM wave, and, as a consequence, the surface currents on the first layer is excited solely by the electric field, it is falsely concluded that the resonance responsible for the perfect absorption is electric one. INstead, the resonance, which induces the perfect absorption, is not an electric one, but a magnetic one. The surface currents in the grounded metallic plane is opposite to those in the first layer for both peaks (see Fig. 3.2c, d of [45]). Therefore, the resonance responsible for the perfect absorption is a magnetic resonance.

[1]The original definition of ELC made by Schurig et al. [46] is the electric inductive-capacitive (LC) resonator with a fundamental mode that couples strongly to a uniform electric field, and negligibly to a uniform magnetic field.

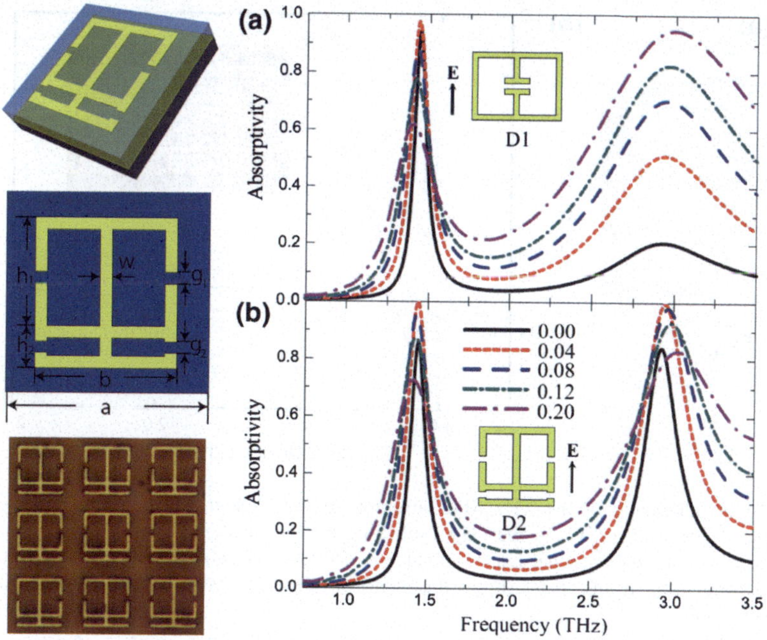

Fig. 3.17 (*Left*) Electric-SRR structures consisting of dual-band resonator and metallic ground plane separated by a dielectric layer. (*Right*) Simulated absorption spectra with different values of tan(δ), the dielectric loss tangent. [44] IOP Publishing. Reproduced with permission. All rights reserved

MMs are usually characterized by an effective complex electric permittivity $\tilde{\varepsilon}_{\rm eff}(\omega) = \varepsilon_1(\omega) + i\varepsilon_2(\omega)$ and an effective complex magnetic permeability $\tilde{\mu}_{\rm eff}(\omega) = \mu_1(\omega) + i\mu_2(\omega)$, which can be obtained by using a concept of effective-medium theory and are definitely different from those of constituent materials. Considerable effort has focused on the real parts of permittivity (ε_1) and permeability (μ_1) to create a negative refractive-index material. To create such structures, it is important to minimize losses (over the operating frequency range) associated with the imaginary parts (ε_2 and μ_2) of the effective response functions. Conversely, for many applications, it would be desirable to maximize the loss. As a consequence, perfect absorption can be realized. Such an absorber, which is compatible with standard microfabrication techniques [47], would be of particular importance at THz frequencies, where it is difficult to find naturally occurring materials with strong absorption coefficients.

Tao et al. [45] experimentally demonstrated an MMPA with an absorption of 0.70 at 1.3 THz. A unit cell of the absorber consists of two distinct metallic elements: an electrical ring resonator and a magnetic resonator, as shown in Fig. 3.18. The electrical ring resonator, which couples strongly with the electric field, consists of two split rings stacked back to back, and negligibly with the magnetic field because of the opposite handedness of the two inductive loops.

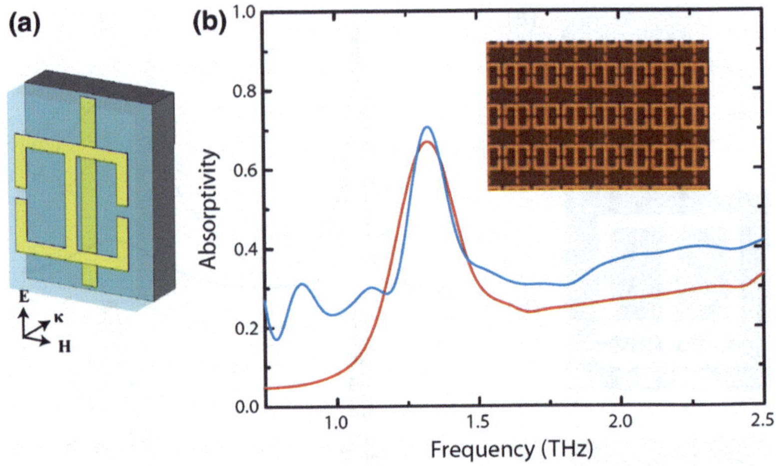

Fig. 3.18 **a** Schematic of the THz MM absorber with the electric resonator on the *top* of a polyimide spacer and a CW on the GaAs substrate for the magnetic coupling. **b** Experimental results showing that the absorber reaches a value of 70 % at 1.3 THz, which is in good agreement with the simulation results. *Inset* shows the photo of fabricated MMPA. Reproduced from [45] with permission

The magnetic coupling is realized by combining the center wire of the electric resonator on the top layer with a cut wire (CW) at the bottom layer in parallel planes, separated by an 8-μm-thick polyimide spacer. The magnetic response can be tuned by changing the geometry of CW and the thickness of polymer spacer. The EM responses, therefore, can be tuned to match the impedance to the free space and to minimize the reflectance at a specific frequency. At 8-μm thickness of MM, this corresponds to a power absorption coefficient of 2000 cm^{-1}, which is significant at THz frequencies, making this low-volume structure a promising candidate for the realization of enhanced, spectrally-selective, thermal detectors, since materials used in current micro-thermal detectors are not fully compatible with micro-fabrication processing, leading to a difficulty in preparation of high-quality thin films and, as a consequence, to a high cost of fabrication.

Shen et al. [48] observed an absorption spectrum (see the top panel of Fig. 3.19) of an MM structure with three near-unity absorption peaks at frequencies of $f_1 = 0.50$, $f_2 = 1.03$ and $f_3 = 1.71$ THz, with absorption of 96.4, 96.3 and 96.7 %, respectively. Furthermore, the simulated absorption peaks, shown in the top panel of Fig. 3.19, are in good agreement with the experimental ones and reach the maximum of 99.5, 99.3 and 99.5 % at frequencies of 0.51, 1.04 and 1.71 THz, respectively. A multiple-reflection interference theory was used to quantitatively analyze the performance of the designed MMPA. An excellent agreement among experimental spectrum, theoretical one obtained by using the multiple-reflection interference theory, and numerically-simulated one, has been achieved. It was also found that, as shown in the bottom panel of Fig. 3.19, the investigated structure may

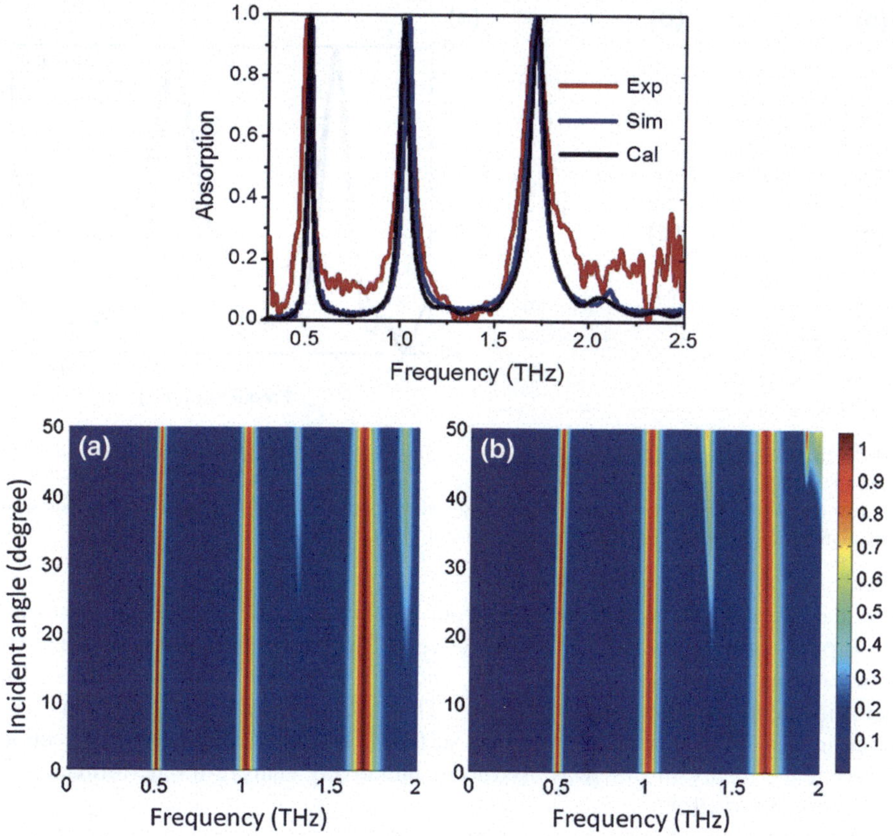

Fig. 3.19 (*Top*) Experimental absorption, full-wave simulated absorption, and theoretically-predicted absorption based on multi-reflection of triple-band THz MM absorber. (*Bottom*) Simulated absorption spectra for **a** TE and **b** TM polarizations, respectively. Reproduced from [48] with permission

be insensitive to the variation of incident angle in a wide range up to 40°. These results are promising for THz absorbing devices, owing to its triple-band absorption in the THz range.

Since the MMPAs fabricated on rigid substrates are limited in applications not only on curved surfaces but also on varying surfaces, ensuring flexibility in MMPAs is one of the challenging tasks for practical applications. Shan et al. [49] demonstrated a flexible dual-band THz MMPA fabricated on a thin polyimide (PI) substrate. The fabricated MMPA is claimed to be 'ultrathin' because the thickness of PI spacer layer is only 25 μm. The MMPA structure shown in Fig. 3.20 consists of periodic array of SRRs with two asymmetric gaps and a metallic grounded plane, separated by a thin-flexible PI spacer. Experimental results show

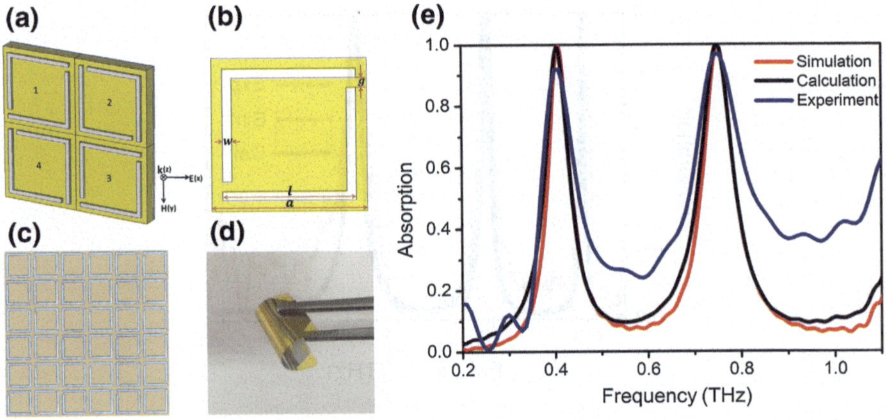

Fig. 3.20 Schematic and absorption spectra of dual-band absorber. **a** Perspective view of unit cell, **b** front view of the single SRR, **c** fabricated structure, **d** photo of thin-flexible dielectric spacer with a layer of Al film and **e** comparison of the absorption spectra obtained by simulation, theoretical calculation, and experiment. Reproduced from [49] with permission

that the absorber has two resonant absorption frequencies (0.41 and 0.75 THz) with absorption of 92.2 and 97.4 %, respectively. They also demonstrated that their MMPA was insensitive to polarization, and had high absorption (over 90 %) for wide incident-angle range from 0° to 45°. The absorption was also better than 90 % even in case of wrapping it to a curved surface (Fig. 3.21). The multiple-reflection interference theory in [48] was used to quantitatively analyze the performance of the designed MMPA.

Apart from the flexibility, tunability is also one of the challenging tasks in MMPAs. Tunable MMPAs can be realized by mechanical reconfiguration, phase change, electrical control, and photo-excitation [50–53]. Recently, Zhao et al. [54] demonstrated highly-flexible and tunable MMPA. The MM unit cells were patterned on thin GaAs patches, which were fashioned in an array on a 10-μm-thick PI substrate via the semiconductor-transfer technique, and the backside of the substrate was coated with gold film as a grounded plane (see the left panel of Fig. 3.22). The thickness of fabricated MMPA was 10 μm, which is only one tenth of wavelength at 1 THz. Optical-pump THz-probe reflection measurements show that the absorption can be tuned up to 25 % at 0.78 THz and 40 % at 1.75 THz through photo-excitation of free carriers in GaAs layers in the presence of 800-nm pump beam (see the right panel of Fig. 3.22). To understand the underlying mechanism of tunability, the authors carried out simulations by assuming that the conductivity of GaAs space layer was varied, originating from the varying density of free carriers induced by the optical pumping.

Instead of the conventional MMPAs with metallic patterns and dielectric spacer, efforts have been made in realizing sophisticated properties with new constituent

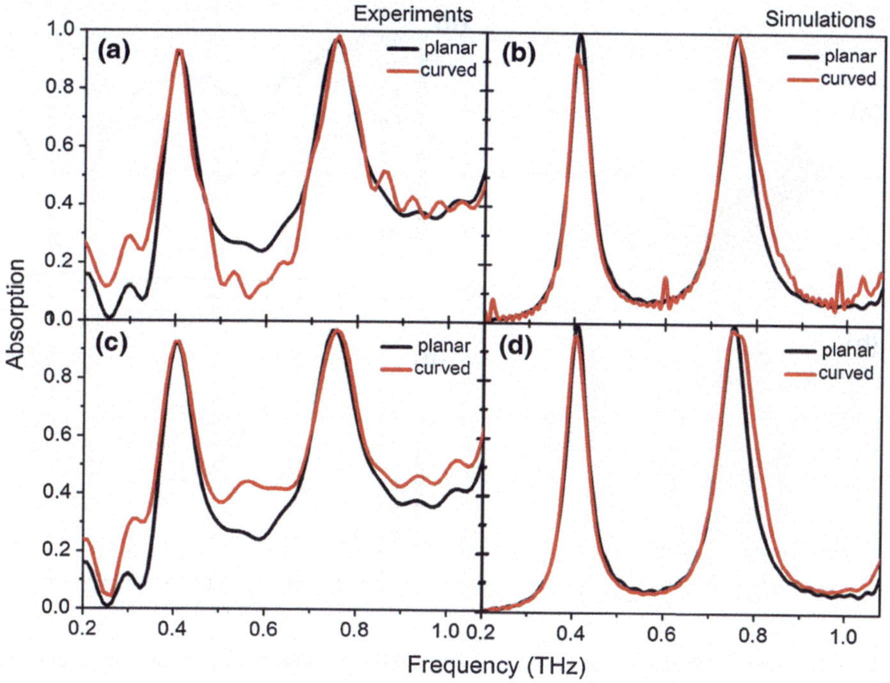

Fig. 3.21 **a** Measured and **b** simulated absorption of planar absorber (*black line*) and curved absorber (*red line*) for TE polarization. **c** Measured and **d** simulated absorption of planar absorber (*black line*) and curved absorber (*red line*) for TM polarization. Reproduced from [49] with permission

materials [55, 56]. Isić et al. [55] proposed an electrically-tunable THz MMPA based on nematic liquid crystals. They showed that a few-μm-thick polarization-independent nematic liquid-crystal MM device displaying THz-reflection modulation depths above 23 dB, ms response times, low operating voltages, and a spectral tuning of more than 15 % (Fig. 3.23). The dramatic improvement of performance is based on invoking the critical coupling with external fields, which rests on a suitable choice of the resonator geometry.

Recent progresses in THZ MMPAs clearly reveal that the future of THz MMPAs is very optimistic, because researchers in this field have exerted enormous efforts in exploiting new application fields and in realizing sophisticated properties of THz MMPAs. It is desirable that in practical application of THz MMPAs the EM-wave response should be broadband, omni-directional, polarization-independent, and flexible. It will be especially challenging if the combination of two or more features, for instance, the combination of omni-directionality and polarization independence, is achieved. In spite of these hardships, multi-dimensional physical properties, such as broadband, flexibility, polarization independence and tunability, can be realized in near future.

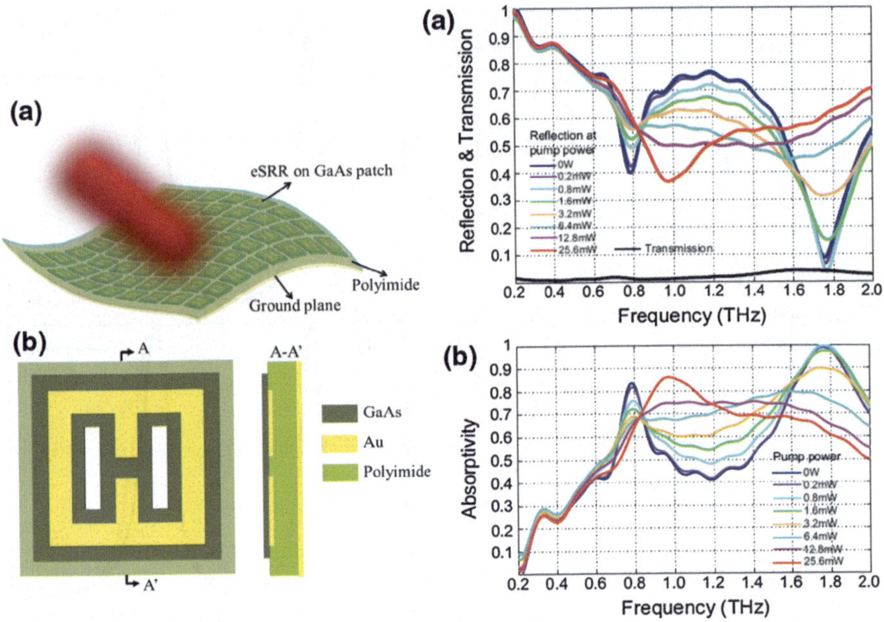

Fig. 3.22 (*Left*) **a** Illustration of flexible tunable MMPA illuminated by an 800-nm beam and **b** unit cell of the MMPA (top view and cross-section view). (*Right*) Experimental results: reflection and transmission coefficients (**a**) and absorption (**b**) at different pump powers. Reproduced from [54] with permission

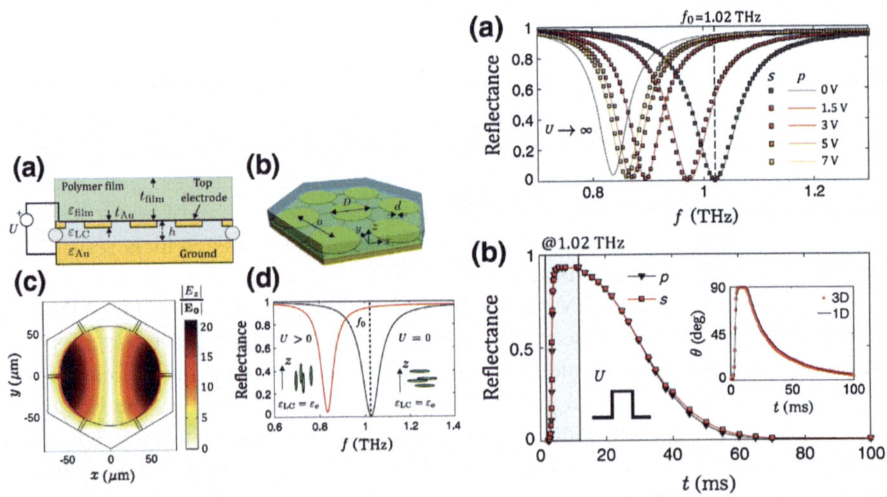

Fig. 3.23 Proposed tunable absorber: **a** cross section (not to scale) and **b** birds-eye view of the hexagonal unit cell and its six nearest neighbors with scalable features. **c** Enhancement of the z component of electric field at the resonant frequency evaluated in the middle of the LC cell and **d** modulation of the reflectance spectra by bias U. (*Right*) **a** Tuning the reflectance spectra with U for s- and p-polarized incident waves. **b** Transient response evaluated at f_0 = 1.02 THz. Reproduced from [55] with permission

3.3 MMPAs Operating in the Infrared and Optical Ranges

It is generally believed that the magnetism is usually associated with radio and lower frequencies, as pointed out by Ozbay [57]. In the optical range, according to Landau (see Chap. 79 of [58]), the magnetism is nearly irrelevant to MMs because magnetic plasma in the optical frequency range is known to be impossible, even though it is very crucial for MMs operating in the infrared (IR) and the optical ranges. As stated by Landau, "there is certainly no meaning in using the magnetic susceptibility from optical frequencies onwards, and in discussion of such phenomena we must put $\mu = \mu_0$." If a particle with a definite spin moment is under an influence of a magnetic field, the direction of spin precesses, which is known as Larmor precession. The frequency of Larmor precession is proportional to the applied field and specific charge of the particle. For example, the electron's specific charge is $e/m_e = 1.76 \times 10^{11}$ C/kg. For a typical strength of magnetic field, 0.01–1 T, the precession frequency is at most a few hundreds of GHz, which is far from the IR and the optical ranges. For a wave with wavelength of 1 μm the frequency is 3×10^{14} Hz. It implies that, therefore, the temporal variation of the magnetic field of EM wave in the IR and the optical ranges is too fast for the electronic spin to respond to the variation of the magnetic field. This is the justification of Landau's statement. Fortunately, Pendry [59] suggested that the SRR structure can be scaled down in order to generate magnetic plasma in optical frequencies.

The most challenging problem in realization of MMPAs operating in the IR and the visible ranges is the size of the unit cell of patterned array. The size of unit cell should be much smaller than the wavelength of incident EM wave to fulfill the deep sub-wavelength condition of MMs. Because the wavelength of IR and optical EM waves is in the range of a few hundreds of nm, the periodicity of the patterned array should be a hundred or a few tens of nm. Current level of techniques of nano-fabrication is sufficient to handle this range of unit-cell size; however, the uniformity of individual unit cells are still challenging. Because of this difficulty progresses in MMPA operating in the IR and the optical ranges were made only recently.

To achieve more practical MMPA, sometimes we need to stack many 'MM/dielectric' bilayers; however, it is very difficult to fulfill the alignment-tolerance requirements [57]. After the first demonstration of a design of MMPA in the GHz range [1] there were many attempts to fabricate MMPA operating in the IR and/or the optical ranges. Avitzour et al. [60] were the first researchers who attempted to design MMPA operating in the IR range. The design of the absorber was based on a perfectly impedance-matched negative-index material. They employed quite complicated structure (see Fig. 2 of [60]). Although the absorption was not very close to unity, it was over 90 %, and the reflection was nearly 0. They claimed that absorption might reach ~99 % if double-layer structure was used. The range of operating angle of incidence was ±45°.

In real application of MMPAs a broadband response is extremely desirable. To accomplish the broadband behavior an approach of multiple-band response is a good way to practice. One way for the multi-band absorption can be achieved by designing the unit cell with multiple resonating components. A highly-efficient conformal dual-band metamaterial was realized by Jiang et al. [61] by using the strategy as follows. The substrate was prepared in two steps. In the first step 50-nm thick Au layer was deposited on Si wafer by using thermal evaporation. The intermediate 100 nm-thick Kapton layer was then deposited on top. The top Au nano-resonator array was fabricated by using electron-beam lithography and a metal lift-off process. The reflection was measured by using a FTIR spectrometer. The unit-cell structure of Au nano-resonator array was optimized by using a genetic algorithm. Figure 3.24 shows the schematic diagram of the optimized structure of unit cell and photo of the fabricated sample with optimum structure. This conformal MMPA maintains its absorption properties when integrated even onto curved surfaces of arbitrary materials. As can be seen in Fig. 3.25a, there is reasonable agreement between simulated and measured reflection spectra. Simulated spectra for both polarizations are identical to each other. The same is true for the measured spectra because of the 8-fold symmetry of the unit cell. The two vertical red strips in Fig. 3.25b clearly show two strong absorption bands nearly independent of the incident angle. By comparing Figs. 3.25b and c we can realize that the two angle-independent absorption bands agree well with the theoretical predictions.

Dayal and Ramakrishna [62] proposed a simple design of MMPA operating in the mid-IR regime with a broadband absorption. In the proposed structure circular metallic disks were arranged periodically, and the metallic-disk layer was separated from a continuous metallic film by a dielectric film. The absorption decreases as the thickness of Au disk decreases, owing to the fact that the charge distribution on the gold disk was not reflected properly in the grounded metal plane due to the finite

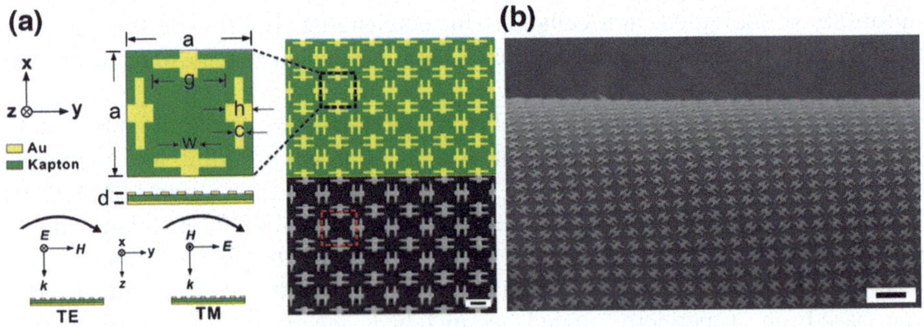

Fig. 3.24 Diagram and FESEM image of the fabricated dual-band mid-IR MMPA. **a** *Top* Doubly-periodic array of H-shaped nano-resonators showing magnified view of one unit cell. *Right* FESEM image of a portion of the fabricated MMPA. Scale = 600 nm. *Bottom left* Orientation of the incident fields with respect to the MMPA. **b** Low-magnification FESEM image of the freestanding fabricated conformal MMPA showing its mechanical flexibility. Scale = 1800 nm. Reproduced from [61] with permission

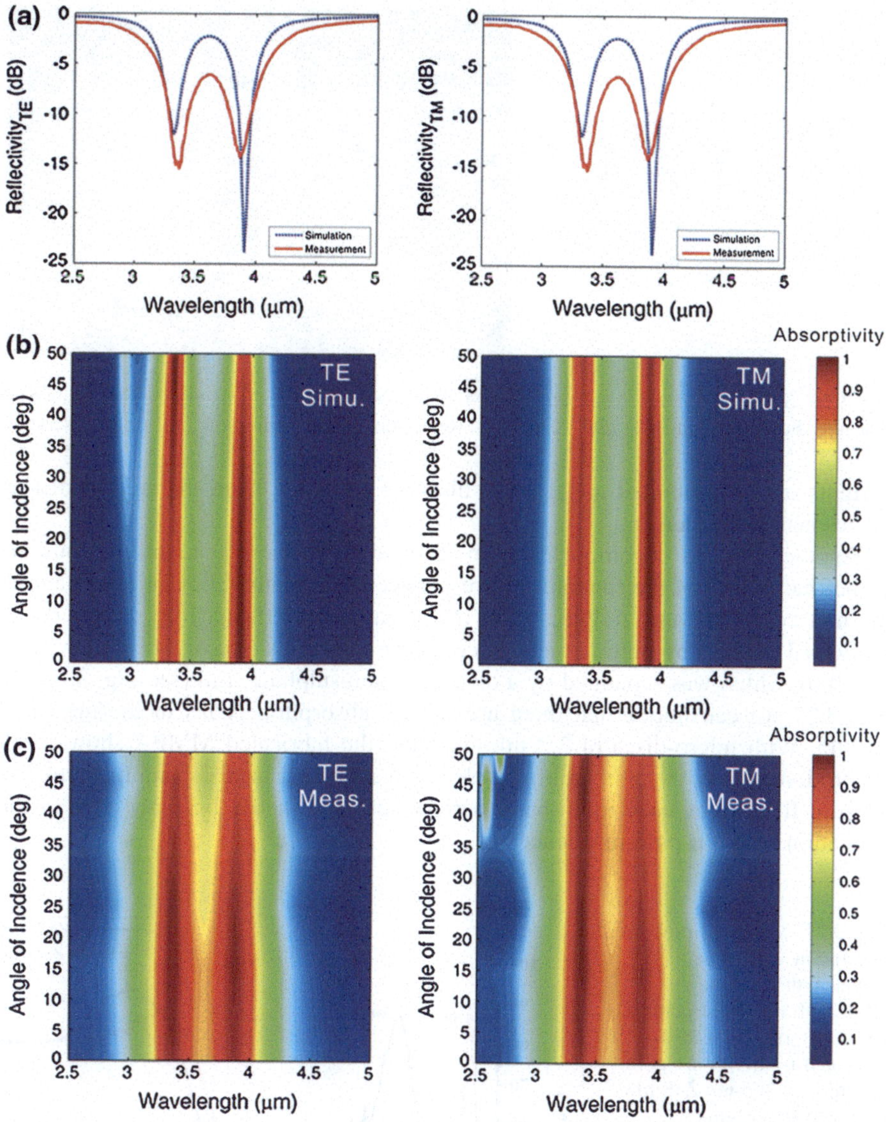

Fig. 3.25 Angular peak dispersion of the absorption spectra. **a** Simulated and measured reflection of the MMPA under TE (*left*) and TM (*right*) illumination at normal incidence. Contour plot of **b** the simulated and **c** the measured absorptivity as a function of wavelength and angle of incidence under TE (*left*) and TM (*right*) illumination. Reproduced from [61] with permission

skin depth. The proposed structure can also operate in a wide range of angle of incidence. The bandwidth of absorption peak was enhanced by putting one more pair of metal/dielectric bilayer on top of the Au disks. The broadband absorption was attributed to a confined surface current in the lower Au disk, and the surface

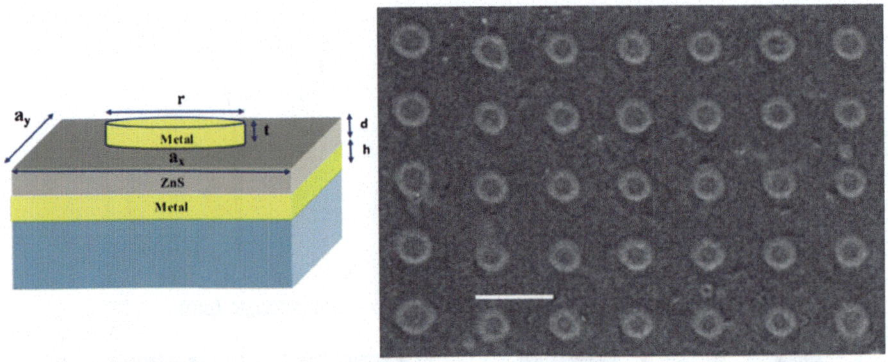

Fig. 3.26 Schematic of a unit cell. *Right* SEM image showing the *top view* of the fabricated MM structure. Scale = 8 μm. [63] IOP Publishing. Reproduced with permission. All rights reserved

currents on the grounded metal plane disperse out away from the area of disk at off-resonance frequencies.

Another way to accomplish the multi-band absorption is to utilize multiple higher harmonics of the fundamental resonance. This method has an advantage of the unit cell with very simple shape. Dayal and Ramakrishna [63] designed and fabricated MMPA by putting metallic circular micro-discs (Au or Al) on a metallic thin film, which was separated by a dielectric zinc-sulphide film (see Fig. 3.26). In Fig. 3.27 we can notice that there are several absorption peaks in the measured spectra. With micro-discs of 3.2 μm diameter, the fabricated MMPA shows peak absorption of over 90 % in multiply-selected wavelength bands spanning from 3 to 14 μm. It behaves as a multi-band perfect absorber in the IR range due to the excitation of multi-pole resonances.

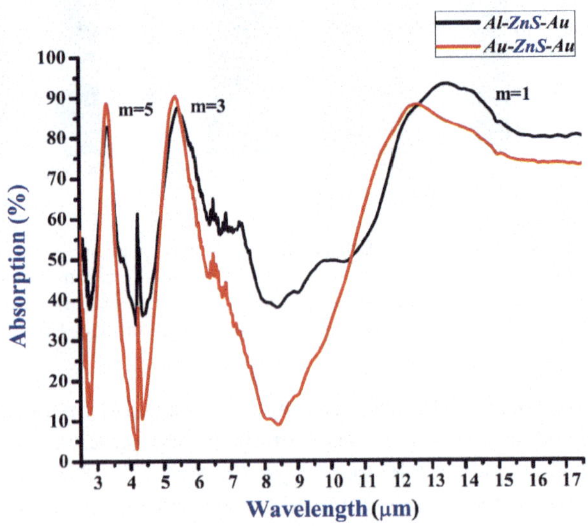

Fig. 3.27 Measured power absorption versus the wavelength from the fabricated MMPA structures consisting of Au/Al discs separated from 100-nm gold thin film by 280-nm ZnS film. [63] IOP Publishing. Reproduced with permission. All rights reserved

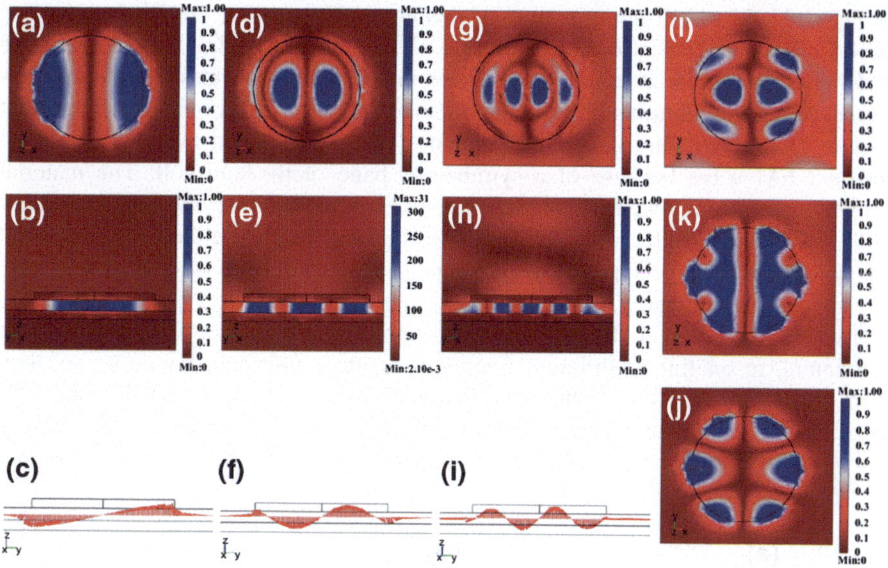

Fig. 3.28 Field distributions for the three modes at resonant wavelengths of 12.58, 5.4 and 3.35 μm for the normal incident angle. **a**, **d** and **g** represent the electric field in the top disk. **b**, **e** and **h** represent the magnitude of magnetic field in tri-layer. **c**, **f** and **i** show the distribution of electric field in the z direction. **j**, **k** and **l** represent the multipolar nature of the electric field in the top disk at 6.16, 5.88 and 4.45 μm, respectively. [63] IOP Publishing. Reproduced with permission.

To understand the detailed mechanism of multiple-band absorption, simulations were performed by using the finite-element method. The simulated absorption spectrum is in a reasonable agreement with the experiment. The main features of results are summarized in Fig. 3.28. Basically there are two groups of resonances. The first group of resonances can be understood by using the antenna theory. The antenna theory in the radio-frequency range predicts that the resonance occurs when the condition $m\lambda/2 = L$, where L is the characteristic length and m is an integer, is satisfied. The condition does not hold anymore in the IR and the optical ranges because the rapid variation of EM field causes the penetration of EM wave into metal, leading to the plasma oscillations. Therefore, at high frequency we have to use different length scale, and the metal respond to a shorter 'effective' wavelength, λ_{eff}, which follows a simple linear scaling [64]

$$\lambda_{\text{eff}} = \alpha_1 + \alpha_2 \frac{\lambda}{\lambda_p}, \tag{3.1}$$

where λ_p is the plasma wavelength, and α_1 and α_2 are coefficients determined by geometry and material properties. By applying this rule the resonances at wavelengths of 12.58, 5.4, and 3.35 μm are the dipolar oscillation and its odd-multiple resonances (see Fig. 3.28a–i). These peaks are designated as $m = 1$, $m = 3$ and $m = 5$

in Fig. 3.27. Most intriguing feature of this article is that there are other resonance peaks at wavelengths of 6.16, 5.88 and 4.45 μm other than the dipolar resonance and the odd-multiple resonances. They are multipolar resonances, which are reminiscences of the vibrating circular membrane (see Fig. 3.28j–l). Another important feature of this structure is the independence of absorption on the polarization of incident EM wave because of a symmetric shape of the unit cell. The material property of metallic patches also plays a minor role. The replacement of Al patches with Au patches results in a slightly inferior absorption, while the overall spectral shape does not change appreciably. Especially, the locations of resonance peaks are not changed significantly.

Recently, Bossard et al. [65] employed a strategy to integrate several nano-scale resonators, resonating at different frequencies, into a unit cell. By doing so, they were successful to realize the broadband absorption in the mid-IR range. In Fig. 3.29 a schematic diagram and the FESEM image of fabricated MMPA is shown. Absorption spectra with different thicknesses of metallic layer are presented

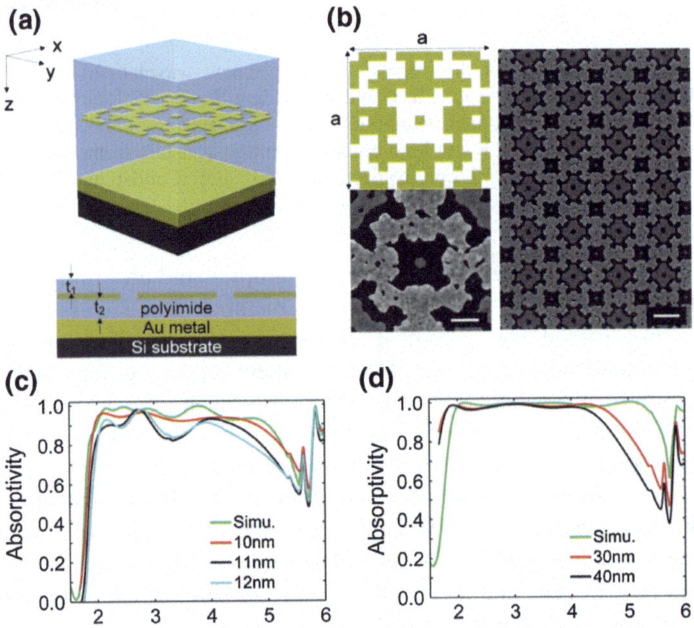

Fig. 3.29 **a** Diagram of the optimized Au-based MMPA structure. **b** *Top left* Top view of one unit cell of the design. *Bottom left* FESEM image of the unit cell of fabricated structure. Scale bar is 200 nm. *Right* Low-magnification FESEM image of the same structure. Scale bar is 600 nm. **c** Simulation and measurement for Au-based MMA under unpolarized illumination at normal incidence. The average Au thicknesses of the Au nanostructures determined by atomic-force-microscopy measurements were 10, 11, and 12 nm. **d** Simulation and measurement for Pd-based MMAs with Pd nanostructure thicknesses of 30 and 40 nm under unpolarized illumination at normal incidence. Reproduced from [65] with permission

in Fig. 3.29c, d. Because of a symmetrical shape of the unit cell, the fabricated MMPA was supposed to be polarization-insensitive and thus the measurements were done with an unpolarized light source. As can be seen in Fig. 3.29c, d, the measured average absorption greater than 98 % was maintained over a wide range of angle of incidence (±45°) for mid-IR wavelengths between 1.77 and 4.81 μm. Because of an integrated nano-resonators operating at different frequencies and doubly periodic structure along the perpendicular direction, many closely-spaced electric resonances can be excited simultaneously, resulting in a broadband-absorption behavior. A genetic algorithm was employed for the optimization of the pattern of unit cell. According to simulations, the authors claimed that large electric currents and field enhancements were induced by the incident EM wave on different parts of the metal nanostructure at wavelengths corresponding to absorption peaks. It implies that the individual elements, which are individually operating at different frequencies, as well as the coupling among them, are important to realize the high absorption.

The dependence of absorption on the polarization of incident EM wave and the angle of incidence is another important issue for practical application of MMPAs, as well as the broadband behavior. The polarization independence can be easily attainable if the shape of unit cell and the metallic patches possess at leat 4-fold symmetry and even higher symmetry. Although the polarization independence can be easily attainable, the realization of incident-angle-insensitive MMPAs is not an easy job, especially, when the polarization independence is required at the same time. For TE mode, in which the electric field is parallel to the reflecting surface, only the tangential and normal components of magnetic field are varying according to the incident angle. Because the magnetic field is interacting with matter weakly compared to the electric field, the TE mode is relatively insensitive to the variation of incident angle. For TM mode, because of varying tangential and normal components of the electric field it is more difficult to attain an incident-angle insensitivity.

Aydin et al. [66] designed, fabricated and characterized a broadband and polarization-independent resonant MMPA by utilizing ultrathin plasmonic super absorbers. First, two grating-type MMPAs were fabricated; one has continuous stripes as ruled grating and the other one has stripes, which is tapered regularly, forming a connected-trapezoid array. Although both structures exhibited rather good absorption properties over the whole visible regime, the trapezoid structure possessed better performance probably because the trapezoid structure had varying width, which led to a broadband response. Based on these results and the idea of symmetric shape of the unit cell and the metallic patch for polarization-independent behavior, a metal-insulator-metal (MIM) stack was fabricated and characterized (see Fig. 3.30a, b). The top layer of the fabricated MMPA is basically a crossed trapezoidal arrays made of Ag. A broadband resonant light absorption over the entire visible spectrum (400–700 nm) was achieved with an average measured absorption of 0.71 and simulated absorption of 0.85. By comparing the right panels of Fig. 3.30 with left panels it can be concluded that the crossed trapezoidal structure shows better performance than a simple fishnet structure. Furthermore, it was, of course, independent of the polarization of incident EM wave (see Fig. 3.30f, g). It was

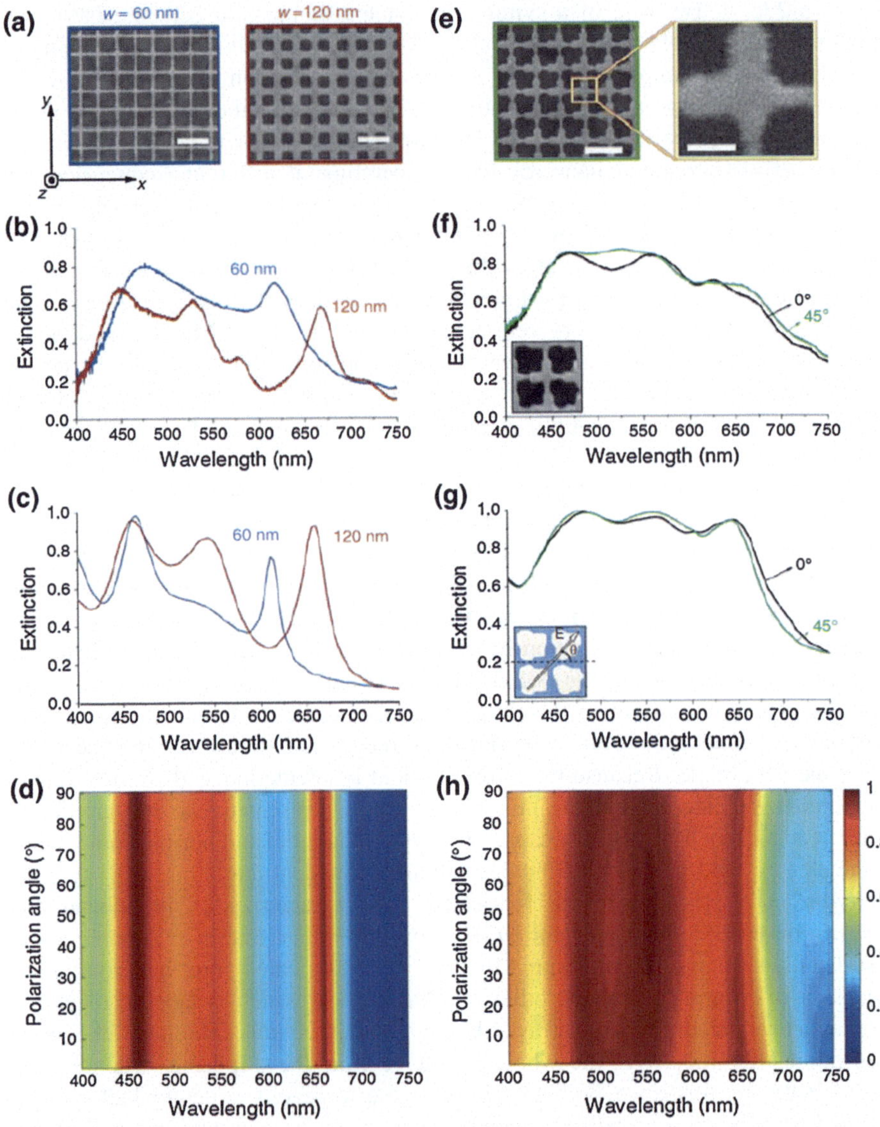

Fig. 3.30 Crossed metallic gratings and broadband polarization-independent plasmonic super absorbers. **a** SEM images of two different crossed metallic gratings. Scale bars are 500 nm. **b** Measured and **c** simulated extinction spectra for two different crossed metallic gratings. **d** Extinction spectra for 120-nm-wide crossed-grating structure as a function of polarization angle and wavelength. **e** SEM image of the fabricated crossed trapezoid arrays (*left*) and a single unit cell of crossed trapezoid (*right*). Scale bars are 500 and 100 nm, respectively. **f** Measured and **g** simulated extinction spectra using digitized SEM images (*Inset*) for different incident polarization angles (*Inset*). **h** Extinction spectra for crossed trapezoid array as a function of polarization angle and wavelength. Reproduced from [66] with permission

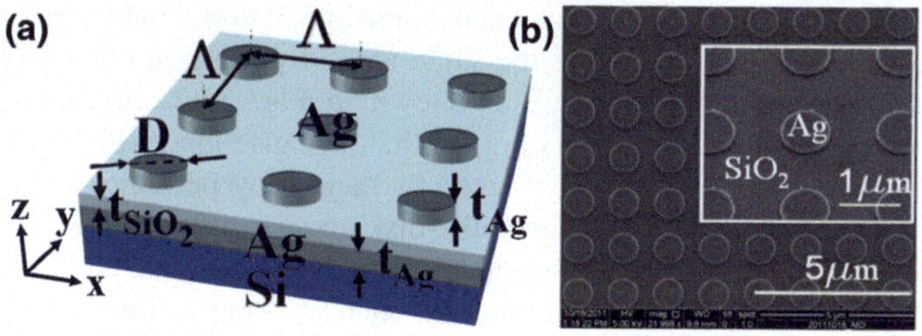

Fig. 3.31 **a** Schematic diagram of the investigated metallic-disk structure. **b** SEM image of the fabricated structure. The *Inset* shows the details of the structure within one unit cell at a 35° angle of view. Reproduced from [67] with permission

suggested that broadband absorption and photocarrier generation can be achieved by replacing the dielectric spacer with an active semiconductor with high absorption coefficient.

A high-performance, wide-angle, polarization-independent dual-band MMPA was realized by manipulating the ratio of disk size to the periodicity and the disk size itself. Cheng et al. [67] designed, fabricated and characterized a dual-band MMPA made of Ag/SiO_2/Ag tri-layer on top of Si wafer (see Fig. 3.31). The top layer is a square array of circular Ag disks. According to simulations it was claimed that the manufactured MMPA had a resonance absorption band around 0.35 eV with absorption better than 84 %, which is persistent in a wide range of incident angle because the metallic disk exhibited a localized surface-plasmon polariton (LSPP) mode for both TE and TM polarizations coupled with Fabry-Pérot-like resonance in the SiO_2 cavity. The absorption spectra were measure by using an FTIR spectrometer. Again, the fabricated MMPA was supposed to be polarization-insensitive and thus the measurements were done with an unpolarized light source. To achieve a broadbandbehavior, the unit cell of top layer is modified to contain two disks with different radii in the unit cell (see Fig. 3.32b, which

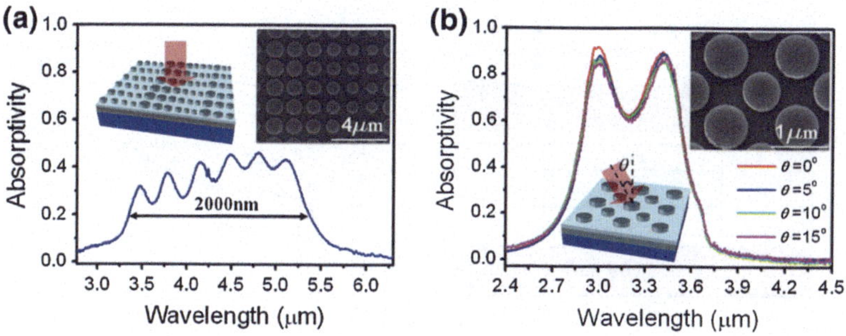

Fig. 3.32 **a** Experimental absorption of the broadband absorber composed of the multi-sized disks. **b** Absorption of the dual-band absorber consisting of two different disk sizes. The *Insets* show the SEM images of one unit cell for the fabricated broadband and dual-band absorbers. Reproduced from [67] with permission

clearly exhibits dual absorption peaks and insensitivity in a wide range of incident angle). Furthermore, by employing the same idea of putting multiple unit patches with different radii in the unit cell, as shown in Fig. 3.32a, a very broadband behavior was achieved. The authors argued that, because the LSPP resonances for the metallic-disk absorbers are not affected by their array periodicity, a broadband emitter composed of a multi-sized metallic-disk array can be realized.

Tunability of MMPA is the most difficult feature to be realized. It is desirable if MM is tunable under an external influence, such as temperature variation, application of electric and/or magnetic field, variation of geometrical parameters, and any other external stimulus, during operation. Werner et al. [68] suggested a tunable MM by using the liquid crystal (LC) as an agent for external stimuli. Schematic of the unit cell of the suggested negative-index material is shown in the left panel of Fig. 3.33. Because the permittivity of LC is given by

$$\varepsilon_{LC} = \frac{\varepsilon_{\parallel}\varepsilon_{\perp}}{\varepsilon_{\parallel}\cos^2\chi + \varepsilon_{\perp}\sin^2\chi}, \tag{3.2}$$

where $\varepsilon_{\perp}(\varepsilon_{\parallel})$ is the permittivity for the incident EM wave polarized perpendicular (parallel) to the director axis of LC and χ is the angle between the director axis and the wave vector of incident EM wave. The direction of the director axis can be adjusted by the external electric field. By varying ε_{LC}, i.e., varying external electric field, the refractive index can vary from negative to zero, and positive, as shown in the right panel of Fig. 3.33. If this MM is realized experimentally, it is, in a real sense, a tunable MM.

Pitchappa et al. [69] reported the tunability of a near-IR micro-electro-mechanical system (MEMS) switchable complementary MMPA. Its schematics of unit cell and switching principles are depicted in Fig. 3.34. The micro-actuator, which will eventually tune the response of MMPA, can be switched on and off by the applied voltage. As shown in Fig. 3.35c, it is evident that the measured absorption characteristic can be tuned. The bias voltage reduces the effective cavity

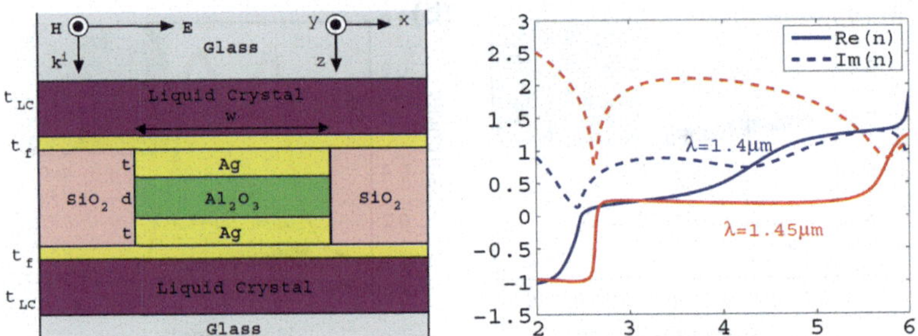

Fig. 3.33 (*Left*) A two-dimensional MM with thick LC superstrate and substrate layers having a reconfigurable index of refraction. (*Right*) Effective index of refraction n with respect to ε_{LC} at two different wavelengths of 1.4 and 1.45 μm. Reproduced from [68] with permission

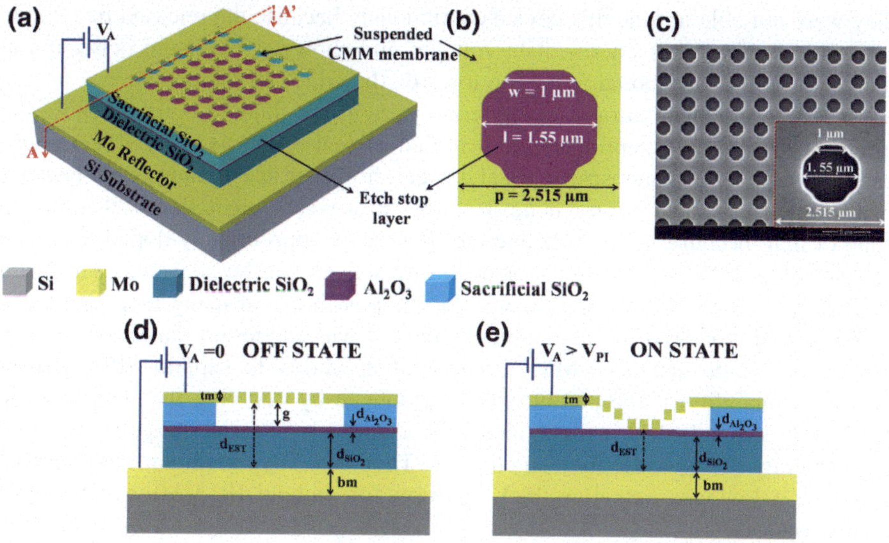

Fig. 3.34 **a** Schematic of near-IR MEMS switchable complementary metamaterial absorber (MSCMA). **b** Schematic top view of the unit cell with the MM geometry definitions. **c** SEM image of the fabricated MSCMA with the *Inset* showing the unit cell definition and **d** and **e** show the schematic cross-sectional view along AA′ of MSCMA in OFF state and ON state, respectively. Reproduced from [69] with permission

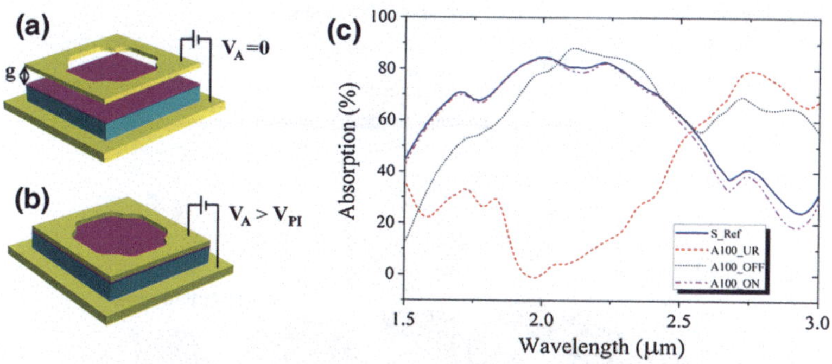

Fig. 3.35 Schematic drawing of **a** released MSCMA in OFF STATE (before actuation voltage is applied, $V_A = 0$ V) and **b** released MSCMA in ON STATE (after actuation voltage is applied, $V_A > V_{PI}$). **c** Measured absorption spectra of S_Ref, A100_UR, A100_OFF, and A100_ON MSCMAs, respectively. Reproduced from [69] with permission

thickness, resulting in a blue shift of the absorption peak. They also fabricated a sample without air gap (see Fig. 3.35a) to verify the ON-state is really gapless. Both measured spectra denoted as S_Ref and A100_ON in Fig. 3.35c are indeed nearly identical. If this device is to be continuously tunable, the air gap should be controlled by bias voltage. Although it was checked by applying different bias voltages, they were not able to tune this device continuously because the released membrane does not remain flat at intermediate voltages and shows a maximum deflection at the center and then gradual reduction towards the edges of membrane.

Recently, graphene attracts many researchers in the field of optics because the graphene possesses exceptional electrical and optical properties. The direct application of graphene in nanoscale optical devices, however, is not an easy job because of the small thickness of graphene, resulting in a weak interaction with light. In spite of that, because in the THz and the IR regions appropriately-doped graphene can generate SPPs, strong light-graphene interactions can be achieved. Vakil and Engheta [70] theoretically investigated for the possibility of using graphene as an optoelectronic material in nasoscopic devices. It was suggested that graphene can replace the noble metals in MMs because of its ability to support SPPs and, in addition, tunability of the permittivity of graphene by applying the static electric field. The applied electric field alters the chemical potential of graphene, and consequently the permittivity also changes. They also suggested that the graphene could be a low-loss one-atom-thick platform for flatland IR MMs and transformation optics.

Based on these results Cai et al. [71] suggested a diffraction-grating-type MMPA utilizing a graphene layer between patterned metallic layer and dielectric layer. Figure 3.36 displays the schematic of investigated MMPA utilizing a nanostructure of metal/dielectric/graphene/dielectric/metal (MDGDM) to enhance and manipulate

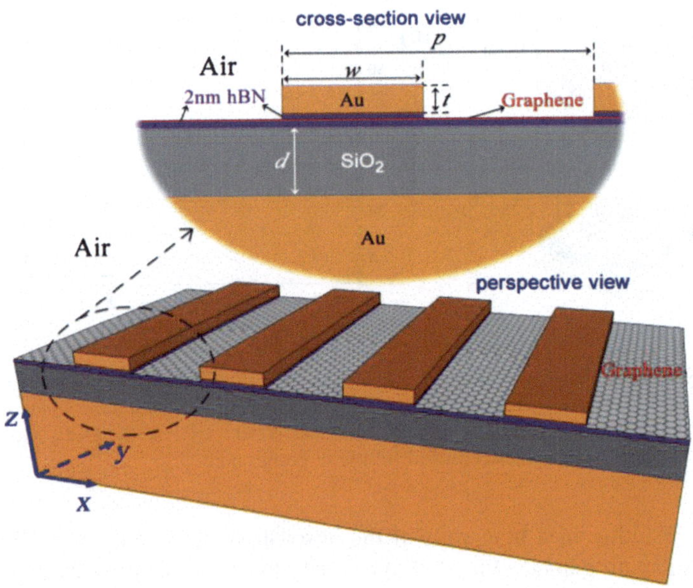

Fig. 3.36 Schematic drawing of an MDGDM nanostructure. *Symbols d* and *p* correspond to the thickness of SiO_2 dielectric layer and the periodic spacing of Au nano-ribbons, respectively. Reproduced from [71] with permission

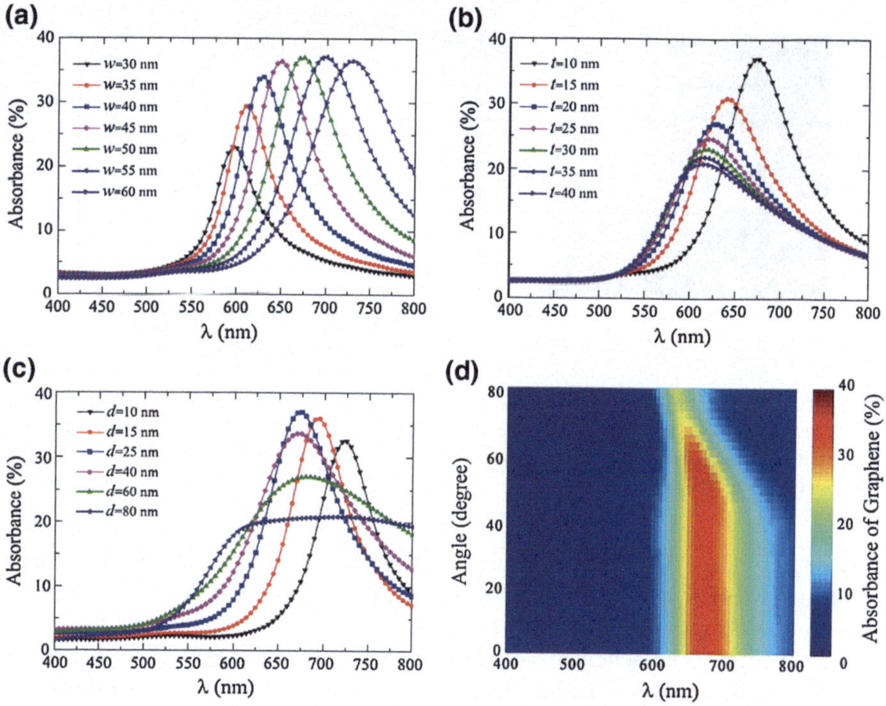

Fig. 3.37 Light absorption of monolayer graphene under normal incident p-polarized light. **a** For different w values, at t = 10 and d = 25 nm. **b** For different t values, at w = 50 and d = 25 nm. **c** Using different d values, for w = 50 nm, t = 10 nm. **d** Absorption of monolayer graphene as a function of wavelength and angle of incidence under p-polarized light. Reproduced from [71] with permission

the light-graphene interaction in the visible region. From Fig. 3.37 we can clearly notice that the absorption spectrum can be modified by varying one of three parameters (w, d and t) and the other two fixed. It was claimed that, based on these results, the designed MDGDM nanostructure is tunable. It is also quite tolerable in a wide range of incident angle.

Nano-composite made of metallic nanoparticles dispersed in a matrix can support the excitation of plasmons due to Mie scattering [72]. Mie scattering comes into play when the particle size is comparable or greater than one tenth of λ where the Rayleigh scattering theory fails. Because the EM fields associated with resonances produced by Mie scattering are confined to volumes much smaller than λ^3, it is called localized surface-plasmon resonances. Nano-composite is ideal for broadband application of MMPA, since the resonance frequency depends on the size of dispersed nano-particles and the size of nanoparticle varies inside the nanocomposite. Furthermore, the resonance frequency also relies on the matrix material. The resonance bandwidth depends on the size, the shape, the density, and the distribution of nanoparticles [72].

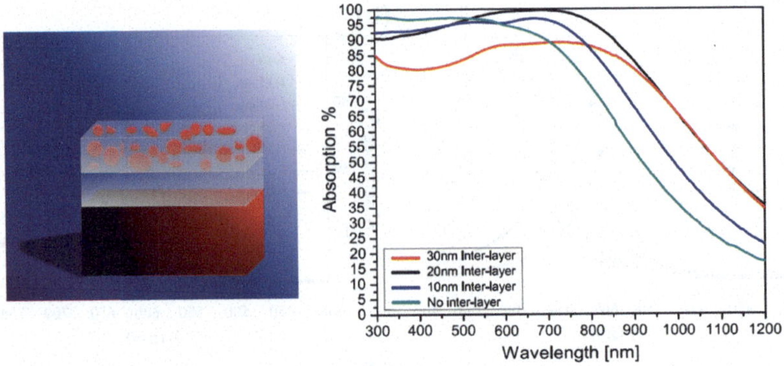

Fig. 3.38 (*Left*) Schematic drawing of the perfect absorber structure, which is composed of the Cu-PTFE nanocomposite deposited on the PTFE coated Cu film. (*Right*) Absorption spectra of 20-nm Cu-PTFE composite on PTFE spacer layer with different thicknesses on a 100-nm Cu. Reproduced from [73] with permission

Hedayati et al. [73] claimed that they were successful to fabricate a tunable broadband MMPA by stacking metal and copper-polytetrafluoroethylene (PTFE) nano-composite. An average absorption of 97.5 % was obtained in the whole visible spectrum. The schematic of MMPA, which is composed of the Cu-PTFE nano-composite deposited on the PTFE-coated Cu film, is shown in the left panel of Fig. 3.38. A sputtering technique was employed for the deposition of metallic and polymeric films. A transmission-electron-microscopy (TEM) image revealed that the deposited polymeric films was the near percolated copper-PTFE nano-composite, and its optical image was completely black compared to the bare Cu plate. As shown in the right panel of Fig. 3.38 20-nm-thick spacer layer of PTFE gives the best performance of device. If the thickness of PTFE spacer layer is thicker or thinner than 20 nmm the absorption was degraded in the whole measured range. It was argued that not only the plasmonic resonances but also the influence of interference affected the absorption. From these results the authors claimed that their MMPA was tunable because the absorption peak could vary depending on the thickness and the type of the spacer layer owing to the sensitivity of plasmon resonance to its environment. Because the fabrication technique is very simple, the production cost is very low compared to the competitive methods, and the fabrication tolerance is very high, this structure can be an outstanding candidate for future application in photovoltaic and sensors.

In [71, 73] it was claimed that their MMPAS were tunable. This claim, however, cannot be regarded as completely true because the tuning cannot be accomplished during operation. Instead, the absorption peak is *pre*-determined in the stage of fabrication. In case of Yao et al. [74], although the similar idea utilized in [71] was employed, a really-tunable MMPA was already realized. Yao et al. fabricated an MMPA schematically depicted in the left panel of Fig. 3.39, and the reflection was obtained according to gate voltage (see the right panel of Fig. 3.39). As can be seen

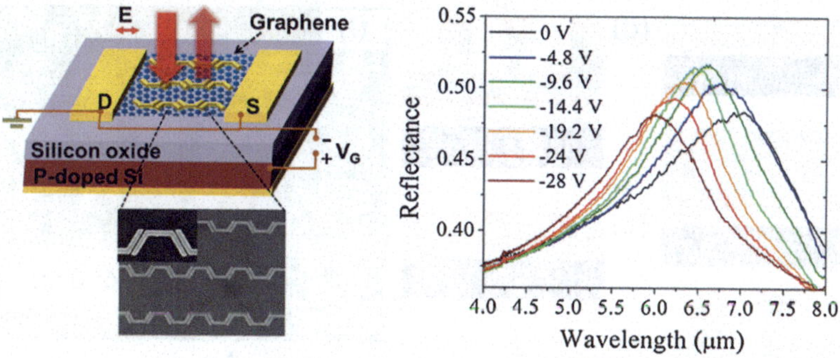

Fig. 3.39 (*Left*) Schematic of the tunable plasmonic device with a back gate. (*Right*) Measured reflection spectra according to gate voltage. Reproduced from [74] with permission

in the left panel of Fig. 3.39, the reflection spectrum can be tuned by the gate voltage, which is applied during the measurements. Therefore, it is called *passive* tuning for the cases of [71, 73], and rest of examples discussed in this section is called *active* tuning.

Flexibility of MMPA is another important issue for practical applications. To make flexible MMPA a flexible material, of course, should be selected for the dielectric spacer. Metals are not very flexible in their bulk form, however, flexibility is not a big problem if they are deposited as thin film. Organic materials possessing ability to conduct electric currents, such as conducting polymers, may replace metals. However, it is not easy to find organic materials which support the excitation of plasmonic resonance. For the dielectric spacer it should possess a durability strong enough for repeated deformations. It seems to be a natural choice of organic materials for the dielectric spacer layer, based on many factors. Moreover, the flexible MMPA should be insensitive to variation of the incident angle because the bent MMPA, in which the incident angle may vary position by position, should work as it were not bent.

Falco et al. [75] fabricated and characterized a flexible MMPA operating in the visible range using Metaflex. Metaflex membranes were fabricated by using a commercially available polymer (SU8, Microchem), which could be spun in thickness varying from a few nm to hundreds of μm. An Au layer of 40-nm thickness was then deposited on top of Metaflex by evaporation. The fabricated MMPA exhibited a plasmonic resonance in the visible region around 620 nm. Authors claimed that one of the most exciting applications of Metaflex is to fabricate 3-D flexible MMs in the optical range, which can be achieved by stacking several Metaflex membranes on top of one another. The natural choice of the pattern of metallic patch is the fishnet. The simulated spectrum agrees with the experimental one reasonably. These results open the door to the generations of 3-D flexible MMs at optical wavelengths.

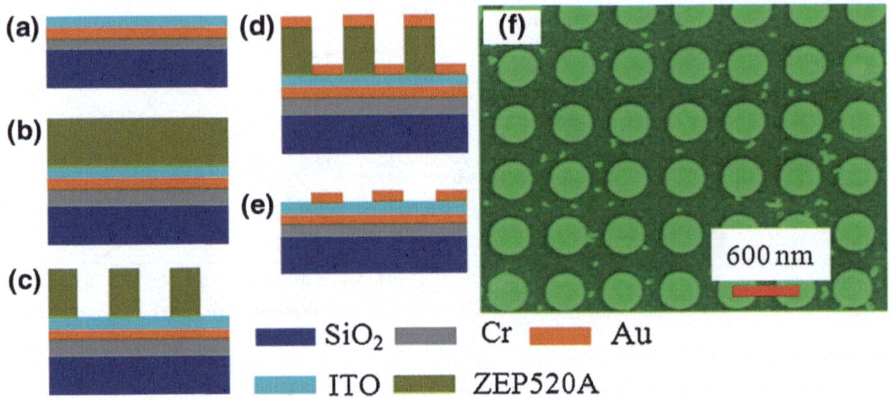

Fig. 3.40 **a–e** *e*-beam lithography steps to fabricate MMPA. **f** SEM image of 2-D Au disc-array MMPA. Reproduced from [76] with permission

A tri-layer (Au-disc/ITO/Au) flexible MMPA working in the near-IR regime, whose schematic diagram is shown in Fig. 3.40, was fabricated on transparent PET substrate by using the flip-chip-transfer (FCT) technique, and characterized [76]. By using the FCT method, the tri-layer Au-disc/ITO/Au absorber MM was transferred from quartz substrate to a transparent PET substrate using optically clear adhesive. Figure 3.41 shows the fabricated flexible MMPA. When the incident angle was increased, the intensity of back reflection became weaker and the absorption dip came to be shallower. This problem should be resolved.

Hedayati et al. [77] studied flexible MMPA employing nano-composite as the top layer. They employed and elaborated the fact that nano-composites with a low-filling factor in the proximity to a thin metallic film, i.e., near the percolation threshold, could enhance the optical transmission of system because of a strong plasmonic coupling between film and nanoparticles, resulting in a new transparent plasmonic device based on a system which is supposed to be a good absorber and scatterer [78]. Hedayati et al. [77] convert this transparent conduction metal (TCM) into broadband MMPA by using TCM for the top layer. The investigated

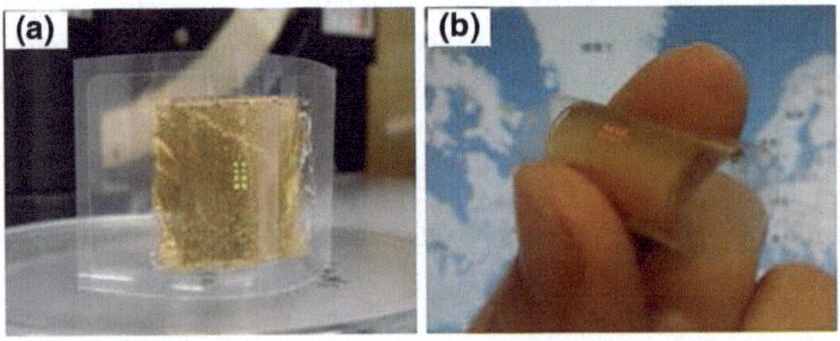

Fig. 3.41 Photos of flexible NIR MMPA. Reproduced from [76] with permission

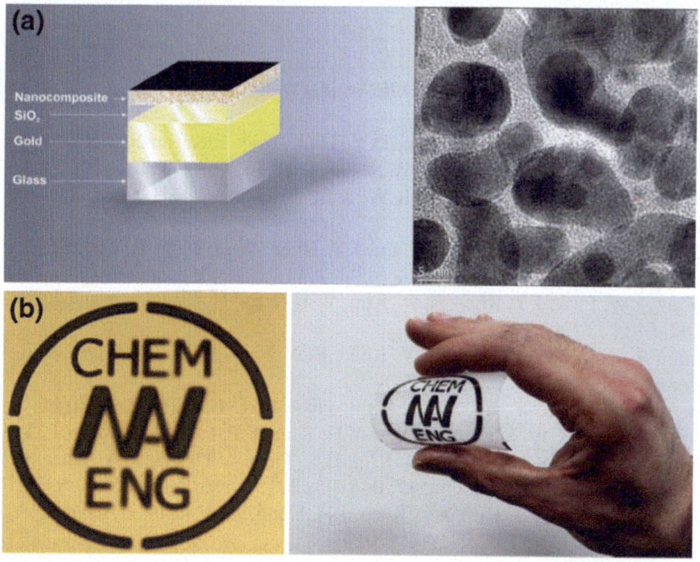

Fig. 3.42 **a** Schematic of MMPA manufactured by sputtering. The *left panel* illustrates that the whole structure resides on a glass substrate; the *right panel* shows a top-view TEM image of the nano-composite film. **b** MMPA (blackbody) coated via a mask on Au-coated glass (*left*) and flexible polymer foil (*right*). Reproduced from [77] with permission

MMPA structure consists of 3 layers: top TCM layer, dielectric spacer layer, and optically-thick metallic film at the bottom. The fabrication was performed by using the sputtering technique with two magnetron sources: an RF magnetron system for sputtering of SiO_2 and a DC magnetron sputter source for Au. The schematic of structure is shown in Fig. 3.42a. By putting a spacer layer, which plays a role of providing a space for transmitted EM wave from the top layer to reside and eventually be absorbed, between the top nano-composite layer and bottom continuous metallic layer, the realization of broadband MMPA was possible (see the left panel of Fig. 3.42b). In addition, flexible MMPA could be fabricated by coating the MMPA structure on a flexible substrate. The variety of nonoparticle size, which supports the excitation of plasmons with different frequencies, leads to a broadband response.

Recent progresses in MMPAs operating in the IR and the optical ranges clearly reveal that in the near future this kind of MMPAs can be realized. Like the case of THz MMPAS, it is desirable that in practical application of this kind of MMPAs, the EM-wave response should be broadband, omni-directional, polarization-independent, and flexible. If this kind of MMPAs is really working in the field, a revolutionary change in field of, for instance, harvesting of energy delivered from the sun may occur.

References

1. N.I. Landy, S. Sajuyigbe, J.J. Mock, D.R. Smith, W.J. Padilla, Phys. Rev. Lett. **100**, 207402 (2008)
2. M. Li, H.L. Yang, X.W. Hou, Y. Tian, D.Y. Hou, Prog. Electromagn. Res. **108**, 37 (2010)
3. X. Shen, T.J. Cui, J. Zhao, H.F. Ma, W.X. Jiang, H. Li, Opt. Express **19**, 9401 (2011)
4. J.W. Park, P. Van Tuong, J.Y. Rhee, K.W. Kim, W.H. Jang, E.H. Choi, L.Y. Chen, Y. Lee, Opt. Express **21**, 9691 (2013)
5. H.Y. Zheng, X.R. Jin, J.W. Park, Y.H. Lu, J.Y. Rhee, W.H. Jang, H. Cheong, Y.P. Lee, Opt. Express **20**, 24002 (2012)
6. P.V. Tuong, J.W. Park, J.Y. Rhee, K.W. Kim, W.H. Jang, H. Cheong, Y.P. Lee, Mater. Chem. Phys. **102**, 081122 (2013)
7. Y.J. Yoo, Y.J. Kim, P. Van Tuong, J.Y. Rhee, K.W. Kim, W.H. Jang, Y.H. Kim, H. Cheong, Y.P. Lee, Opt. Express **21**, 32484 (2013)
8. F. Ding, Y. Cui, X. Ge, Y. Jin, S. He, Appl. Phys. Lett. **100**, 103506 (2012)
9. Y.J. Kim, Y.J. Yoo, K.W. Kim, J.Y. Rhee, Y.H. Kim, Y.P. Lee, Opt. Express **23**, 3861 (2015)
10. N.V. Dung, P.V. Tuong, Y.J. Yoo, Y.J. Kim, B.S. Tung, V.D. Lam, J.Y. Rhee, K.W. Kim, Y. H. Kim, L.Y. Chen, Y.P. Lee, J. Opt. **17**, 045105 (2015). doi:10.1088/2040-8978/17/4/045105
11. Y.J. Yoo, H.Y. Zheng, Y.J. Kim, J.Y. Rhee, J.H. Kang, K.W. Kim, H. Cheong, Y.H. Kim, Y. P. Lee, Appl. Phys. Lett. **105**, 041902 (2014)
12. B.X. Khuyen, B.S. Tung, N.V. Dung, Y.J. Yoo, Y.J. Kim, K.W. Kim, V.D. Lam, J.G. Yang, Y.P Lee, J. Appl. Phys. **117**, 243105 (2015)
13. M. Tonouchi, Nature Photon. **1**, 97 (2007)
14. G.P. Williams, Rep. Progr. Phys. **69**, 301 (2006)
15. S. Borri, P. Patimisco, A. Sampaolo, H.E. Beere, D.A. Ritchie, M.S. Vitiello, G. Scamarcio, V. Spagnolo, Appl. Phys. Lett. **103**, 02110 (2013)
16. H.T. Chen, W.J. Padilla, J.M.O. Zide, A.C. Gossard, A.J. Taylor, R.D. Averitt, Nature **444**, 597 (2006)
17. J. Ma, Y. Zeng, X. Cao, in *Automation Congress, 2008. WAC 2008. World* (2008), p. 11661169
18. J.B. Pendry, A.J. Holden, D.J. Robbins, W.J. Stewart, J. Phys.: Condens. Matter **10**, 4785 (1998)
19. M. Wiltshire, in *Metamaterials and Plasmonics: Fundamentals, Modelling, Applications*, ed. by S. Zouhdi, A. Sihvola, A.P.E. Vinogradov (Springer-Verlag, Marrakech (Morocco), 2009), pp. 191–200
20. M.C.K. Wiltshire, J.B. Pendry, J.V. Hajnal, J. Phys.: Condens. Matter **21**, 292201 (2009)
21. G. Dolling, C. Enkrich, M. Wegener, J.F. Zhou, C.M. Soukoulis, S. Linden, Opt. Lett. **30**, 3198 (2005)
22. C. Imhof, R. Zengerle, Opt. Express **14**, 8257 (2006)
23. D.A. Powell, I.V. Shadrivov, Y.S. Kivshar, Opt. Express **16**, 15185 (2008)
24. G. Dolling, M.W. Klein, M. Wegener, A. Schädle, B. Kettner, S. Burger, S. Linden, Opt. Express **15**, 14219 (2007)
25. J. Valentine, S. Zhang, T. Zentgraf, E. Ulin-Avila, D.A. Genov, G. Bartal, X. Zhang, Nature **455**, 376 (2008)
26. J. Zhou, T. Koschny, M. Kafesaki, C.M. Soukoulis, Phys. Rev. B **80**, 035109 (2009)
27. H. Guo, N. Liu, L. Fu, T.P. Meyrath, T. Zentgraf, H. Schweizer, H. Giessen, Opt. Express **15**, 12095 (2007)
28. D. Li, Y. Xie, J. Zhang, J. Li, Z. Chen, J. Electromagn. Waves Appl. **22**, 1420 (2008)
29. J. Wang, S. Qu, Z. Xu, J. Zhang, H. Ma, Y. Yang, C. Gu, Photon. Nanostruct.: Fund. Appl. **7**, 108 (2009)
30. Z. Li, Y. Ma, R. Huang, R. Singh, J. Gu, Z. Tian, J. Han, W. Zhang, Opt. Express **19**, 8912 (2011)

31. G. Kenanakis, R. Zhao, N. Katsarakis, M. Kafesaki, C.M. Soukoulis, E.N. Economou, Opt. Express **22**, 12149 (2014)
32. N. Liu, L. Langguth, T. Weiss, J. Kästel, M. Fleischhauer, T. Pfau, H. Giessen, Nature Mater. **8**, 758 (2009)
33. C.Y. Chen, I.W. Un, N.H. Tai, T.J. Yen, Opt. Express **17**, 15372 (2009)
34. Christ, O.J.F. Martin, Y. Ekinci, N.A. Gippius, S.G. Tikhodeev, Nano Lett. **8**, 2171 (2008)
35. H.T. Chen, J.F. O'Hara, A.J. Taylor, R.D. Averitt, C. Highstrete, M. Lee, W.J. Padilla, Opt. Express **15**, 1084 (2007)
36. T. Driscoll, G.O. Andreev, D.N. Basov, S. Palit, S.Y. Cho, N.M. Jokerst, D.R. Smith, Appl. Phys. Lett. **91**, 062511 (2007)
37. E. Ekmekci, K. Topalli, T. Akin, G. Turhan-Sayan, Opt. Express **17**, 16046 (2009)
38. J.F. O'Hara, R. Singh, I. Brener, E. Smirnova, J. Han, A.J. Taylor, W. Zhang, Opt. Express **16**, 1786 (2008)
39. R. Singh, C. Rockstuhl, F. Lederer, W. Zhang, Phys. Rev. B **79**, 085111 (2009)
40. Q. Bai, C. Liu, J. Chen, C. Cheng, M. Kang, H.T. Wang, J. Appl. Phys. **107**, 093104 (2010)
41. T.J. Yen, W.J. Padilla, N. Fang, D.C. Vier, D.R. Smith, J.B. Pendry, D.N. Basov, X. Zhang, Science **303**, 1494 (2004)
42. J.Y. Rhee, Y.J. Yoo, K.W. Kim, Y.J. Kim, Y.P. Lee, J. Electromagn. Waves Appl. **28**, 1541 (2014)
43. Z.J. Wang, S.H. Guo, J.J. Li, F.Y. Wang, J. Chem. Pharm. Res. **6**(4), 96 (2014)
44. H. Tao, C.M. Bingham, D. Pilon, K. Fan, A.C. Strikwerda, D. Shrekenhamer, W.J. Padilla, X. Zhang, R.D. Averitt, J. Phys. D. Appl. Phys. **43**, 225102 (2010). doi:10.1088/0022-3727/43/22/225102
45. H. Tao, N.I. Landy, C.M. Bingham, X. Zhang, R.D. Averitt, W.J. Padilla, Opt. Express **16**, 7181 (2008)
46. D. Schurig, J.J. Mock, D.R. Smith, Appl. Phys. Lett. **88**, 041109 (2006)
47. H. Tao, W.J. Padilla, X. Zhang, R.D. Averitt, IEEE. J. Sel. Top. Quant. Electron. **17**, 92 (2011)
48. X. Shen, Y. Yang, Y. Zang, J. Gu, J. Han, W. Zhang, T. Jun Cui, Appl. Phys. Lett. **101**, 154102 (2012)
49. Y. Shan, L. Chen, C. Shi, Z. Cheng, X. Zang, B. Xu, Y. Zhu, Opt. Commun. **350**, 63 (2015)
50. X. Liu, W.J. Padilla, Adv. Opt. Mater. **1**, 559 (2013)
51. T. Driscoll, S. Palit, M.M. Qazilbash, M. Brehm, F. Keilmann, B.G. Chae, S.J. Yun, H.T. Kim, S.Y. Cho, N.M. Jokerst, D.R. Smith, D.N. Basov, Appl. Phys. Lett. **93**, 024101 (2008)
52. D. Shrekenhamer, W.C. Chen, W.J. Padilla, Phys. Rev. Lett. **110**, 177403 (2013)
53. K. Fan, A.C. Strikwerda, X. Zhang, R.D. Averitt, Phys. Rev. B **87**, 161104 (2013)
54. X. Zhao, K. Fan, J. Zhang, H.R. Seren, G.D. Metcalfe, M. Wraback, R.D. Averitt, X. Zhang, Sensor. Actuat. A: Phys. **231**, 74 (2015)
55. G. Isić, B. Vasić, D.C. Zografopoulos, R. Beccherelli, R.c.v. Gajić, Phys. Rev. Appl. **3**, 064007 (2015)
56. B. Zhang, Q. Wu, C. Pan, R. Feng, J. Xu, C. Lou, X. Wang, F. Yang, Opt. Express **23**, 16042 (2015)
57. E. Ozbay, Opt. Photon. News **19**, 22 (2008)
58. L.D. Landau, E.M. Liftshitz, L.P. Pitaevskii, *Electrodynamics of Continuous Media,* 2nd ed. (Pergamon, 1984)
59. J.B. Pendry, Phys. Rev. Lett. **85**, 3966 (2000)
60. Y. Avitzour, Y.A. Urzhumov, G. Shvets, Phys. Rev. B **79**, 045131 (2009)
61. Z.H. Jiang, S. Yun, F. Toor, D.H. Werner, T.S. Mayer, ACS Nano **5**, 4641 (2011)
62. G. Dayal, S.A. Ramakrishna, Opt. Express **20**, 17503 (2012)
63. G. Dayal, S.A. Ramakrishna, J. Opt. **16**, 094016 (2014). doi:10.1088/2040-8978/16/9/094016
64. L. Novotny, Phys. Rev. Lett. **98**, 266802 (2007)
65. J.A. Bossard, L. Lin, S. Yun, L. Liu, D.H. Werner, T.S. Mayer, ACS Nano **8**, 1517 (2014)
66. K. Aydin, V.E. Ferry, R.M. Briggs, H.A. Atwater, Nature Commun. **2**, 517 (2011)

67. C.W. Cheng, M.N. Abbas, C.W. Chiu, K.T. Lai, M.H. Shih, Y.C. Chang, Opt. Express **20**, 10376 (2012)
68. D.H. Werner, D.H. Kwon, I.C. Khoo, A.V. Kildishev, V.M. Shalaev, Opt. Express **15**, 3342 (2007)
69. P. Pitchappa, C. Pei Ho, P. Kropelnicki, N. Singh, D.L. Kwong, C. Lee, Appl. Phys. Lett. **104**, 201114 (2014)
70. A. Vakil, N. Engheta, Science **332**, 1291 (2011)
71. Y. Cai, J. Zhu, Q.H. Liu, Appl. Phys. Lett. **106**, 043105 (2015)
72. W.A. Murray, W.L. Barnes, Adv. Mater. **19**, 3771 (2007)
73. M. Hedayati, F. Faupel, M. Elbahri, Appl. Phys. A **109**, 769 (2012)
74. Y. Yao, M.A. Kats, R. Shankar, Y. Song, J. Kong, M. Loncar, F. Capasso, Nano Lett. **14**, 214 (2014)
75. A. Di Falco, M. Ploschner, T.F. Krauss, New J. Phys. **12**, 113006 (2010)
76. G.X. Li, S.M. Chen, W.H. Wong, E.Y.B. Pun, K.W. Cheah, Opt. Express **20**, 397 (2012)
77. M.K. Hedayati, M. Javaherirahim, B. Mozooni, R. Abdelaziz, A. Tavassolizadeh, V.S.K. Chakravadhanula, V. Zaporojtchenko, T. Strunkus, F. Faupel, M. Elbahri, Adv. Mater. **23**, 5410 (2011)
78. M. Elbahri, M.K. Hedayati, V.S. Kiran Chakravadhanula, M. Jamali, T. Strunkus, V. Zaporojtchenko, F. Faupel, Adv. Mater. **23**, 1993 (2011)

Chapter 4
MMPA, Based on Electromagnetically-Induced Transparency

Abstract Interests in asymmetric metamaterials (MMs) are increasing due to the existence of fascinating phenomena, such as high-quality-factor Fano resonance, analog of electromagnetically-induced transparency (EIT), slow light and chirality. However, there have been few researches studying the application of them to MM perfect absorbers. In this chapter, we discuss about the exploitation of EIT effect and the asymmetric resonators in achieving MM perfect absorbers. By considering the phase coupling between resonators, the EIT can be transformed into the electromagnetically-induced absorption, giving rise to an ultra-narrow and high quality-factor absorption. In addition, breaking the symmetry of resonators can induce multi-resonance that is suitable for developing multi-band or broadband absorption.

4.1 Introduction

Basically, electromagnetically-induced transparency (EIT) is a quantum phenomenon caused by the interference between different excitation sources that makes a medium transparent to the incident electromagnetic (EM) wave [1–3]. Unfortunately, the quantum EIT phenomenon requires rigorous conditions which are normally achievable only in laboratories. Recently, EIT has been mimicked and easily achieved under normal conditions, but classically by using MMs [4–15]. Therefore, real applications in various devices, such as sensors [16–18] and slow-light devices [19, 20], come to be more realistic and can be accomplished in the near future.

So far, two preferential approaches have been used to produce the EIT in MMs (EIT-MMs). The first type is called as bright-bright coupling, where both resonances can be directly excited by the incident EM wave. Exploiting this coupling, Fedotov et al. demonstrated a typical EIT-MM with a high quality factor in the microwave range [21]. By breaking the symmetry of MM structure, which consists of two split-ring resonators (SRRs), anti-symmetric currents were generated and

Y.P. Lee et al., *Metamaterials for Perfect Absorption*,
Springer Series in Materials Science 236, DOI 10.1007/978-981-10-0105-5_4

emitted fields which interfere destructively. Consequently, the incident EM wave propagated without loss, leading to a narrow transmission region in the spectrum. The second type is referred as bright-dark coupling, where only one resonance can be excited by the EM field and the other is stimulated by the near-field coupling of the initial resonance [4, 13, 22, 23]. Therefore, the former and the latter resonances are termed bright and dark modes, respectively. Although dark modes often exhibit higher quality factor of resonance than bright modes, both bright-bright coupling and bright-dark coupling can induce a highly dispersive transparency.

The EM properties of EIT-MMs were described by a common model, the so-called coupled Lorentz-oscillator model [4, 6]. However, only the absorption property was quantitatively obtained while the other scattering parameters, such as reflection and transmission, could not be modeled. Recently, Tassin et al. have presented a sophisticated model which expressed effectively all scattering parameters [24]. Therefore, the group delay, an important factor of slow-light devices, could be determined.

One of the early EIT-MMs was theoretically investigated by Zhang et al. [4]. The authors proposed a structure whose unit cell consisted of a vertical wire and two horizontal wires, being parallel and perpendicular to the electric field, respectively (Fig. 4.1a). The vertical wire was a bright meta-atom, while two horizontal wires played the role as a dark meta-atom. There were two different

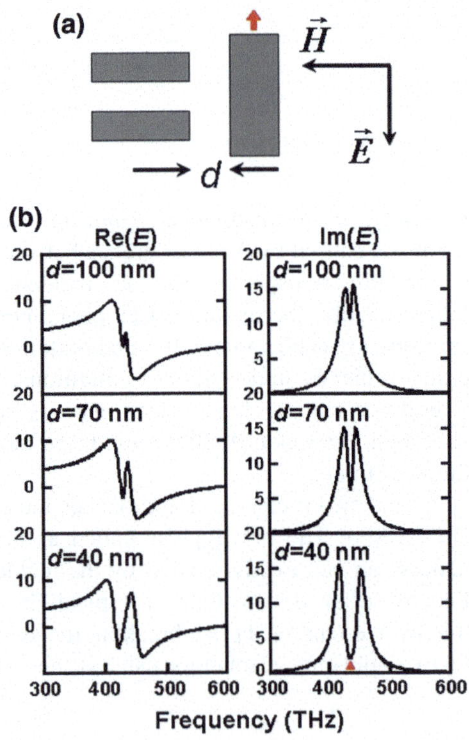

Fig. 4.1 **a** Plasmon-induced transparency MM. **b** Real and imaginary part of E_x probe, placed at 10 nm from the end facet of the radiative antenna [*red arrow* in (**a**)] [4]

pathways of excitation produced by this configuration. The former was a direct excitation of bright meta-atom by incident EM wave. The latter was an indirect excitation of dark meta-atom by the bright meta-atom and, then, couples back to the bright meta-atom. The destructive interference between two excitation pathways canceled the field radiated by the bright meta-atom, leading that the EIT-MM became transparent to the EM wave (Fig. 4.1b).

Subsequently, many configurations of EIT-MMs using the same mechanism have been demonstrated in the microwave [9], the THz [8, 10, 12, 22, 23, 25], and the optical regime [26, 27]. Most of the aforementioned schemes employed planar configurations where only near-field coupling occurred. Recently, a new configuration of EIT-MMs has been considered, which introduced not only near-field but also far-field coupling [28]. In this work, Taubert and coworkers proposed a structure (Fig. 4.2b, c) which was quite similar to the aforementioned structure by Zhang et al. [4]. In place of using the planar configuration, the vertical wire was stacked on top of the horizontal wires. They suggested that the total effect was the sum of the near-field coupling, which generated a quadrupolar oscillation (dark mode), and the far-field coupling, which gave rise to the retardation effect (Fig. 4.2a). By exploiting these effects, a new phenomenon, the so-called electromagnetically-induced absorption (EIA), was observed, where an enhancement of absorption occurred instead of transmission (Fig. 4.2d). This discovery provided interesting

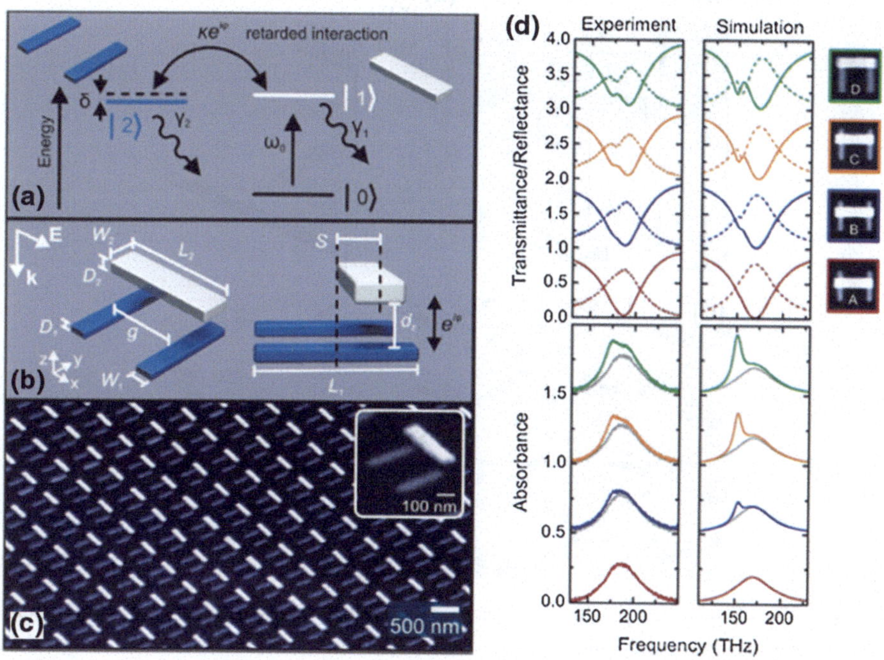

Fig. 4.2 Electromagnetically-induced absorption MM. **a** Three-level scheme for the MM. **b** MM structure. **c** MM sample. **d** Simulated and measured transmission, reflection and absorption spectra according to the displacement of dipole wire S = 30, 45, 60, and 150 nm [28]

capability of phase manipulation in plasmonic-coupling devices. However, the absorption of this proposed structure was only about 50 %, and the EIA-like MMPA needs to be further elucidated.

4.2 Narrow Perfect Absorbers Based on Plasmonic Analog of Electromagnetically-Induced Absorption

Based on previous study on the EIA effect, an ultra-narrow band perfect absorber has been investigated recently [29]. The proposed EIA structure, which consisted of a long slot in a continuous metal film and a moveable short metallic bar, is illustrated in Fig. 4.3. According to the Babinet's principle, the role of electric and

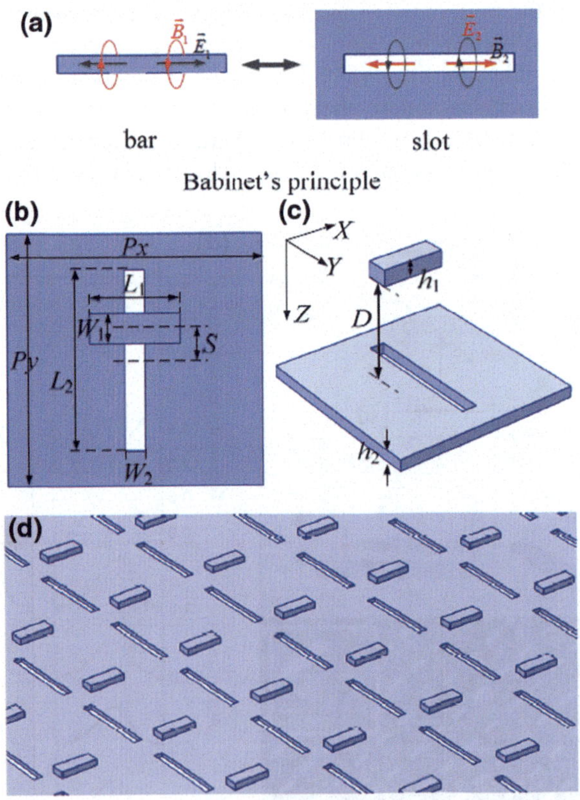

Fig. 4.3 **a** Description by Babinet's principle for the electric quadrupole in a resonator (a metal bar) and the magnetic quadrupole in its complement (a slot in a continuous metal film). **b** and **c** Schematic of the proposed EIA structure with the defined geometrical parameters: $L_1 \times W_1 \times h_1 = 168 \times 60 \times 30\ \mathrm{nm}^3$, and $L_2 \times W_2 \times h_2 = 360 \times 40 \times 20\ \mathrm{nm}^3$. The periods along X- and Y-axis were 500 nm. **d** Oblique view of the stacked MM. The periods along X- and Y-axis were 500 nm [29]

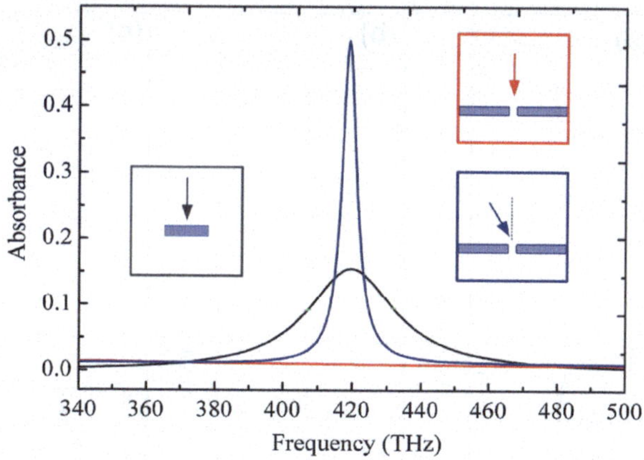

Fig. 4.4 Absorption spectra of the bare-bar array at normal incidence (*black*), and the bare-slot array at normal (*red*) and 20°-off-normal incidence (*blue*). *Insets* Illustration of excitation ways in the *X*-*Z* section. The plane wave was polarized along *X*-axis [29]

magnetic field in a plasmonic structure and its complement can be exchanged. Therefore, if a metal bar can induce an electric quadrupole, then a magnetic quarupole can be excited in a slot. In this structure, the short metal bar can support a radiative dipole mode (bright mode), which can be directly excited by the incident filed. The slot behaves as a non-radiative magnetic quarupole (dark mode), which is only excited via the coupling of the bright mode consisting of both near- and far-field coupling.

The role of the bar and the slot were verified, as shown in Fig. 4.4. The bar induced a broad-linewidth dipole resonance with a rather low absorption. In contrast, there was no response from the slot at normal incidence. However, when the incident angle was 20°, a sharp resonance with comparatively high absorption was induced at the same frequency as that of the dipole resonance excited by the bar. This indicated that a magnetic quarupole mode (dark mode) was excited in the slot by phase retardation, which was unavailable at normal incidence.

Figure 4.5a, b show the evolution of transmission, reflection and absorption spectra of the EIA MM. For the case of symmetry, there was only a broad absorption corresponding to the dipole resonance of the bar. When the symmetry was broken by displacing the bar, a sharp EIA-like absorption peak, which was induced by the coupling of the bar, arose at the center frequency of the broad absorption. The absorption was enhanced by increasing the displacement of the bar, and it reached a nearly perfect absorption of over 97 % with an extremely narrow linewidth (the FWHM was approximately 7.5 nm) when $S = 70$ nm. This absorption peak was 4.85 times stronger in comparison with the symmetric case. Nevertheless, further displacement of the bar decreased the absorption of EIA structure.

To elucidate the nearly perfect absorption in the EIA structure, a similar structure, where the slot in metal film was replaced by a long metal bar, was studied

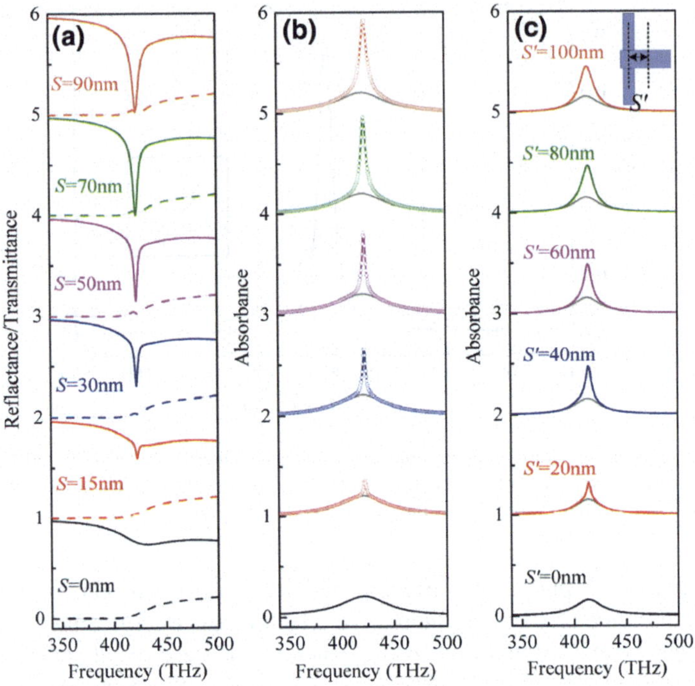

Fig. 4.5 Simulated **a** transmission/reflection (*dashed/solid*) and **b** absorption spectra of the proposed EIA MM according to displacement *S*. The *solid circles* in (**b**) represent the two-coupled-oscillators (TCO)-fit results. For comparison, the absorption spectra of other similar structure by replacing the bottom slot with a long metal bar, in dependence on parameter S', are given in (**c**). *Inset* Top view of the structure with the defined S'. To guide the eyes, the absorption spectra for $S = 0$ and $S' = 0$ (*light gray lines*) are shown in every plot of (**b**) and (**c**), respectively [29]

(Fig. 4.5c). For the optimization of EIA effect, the geometrical parameters were the same except $D' = 150$ nm and $L_2' \times W_2' \times h_2' = 330 \times 40 \times 20$ nm^3. In this comparison structure, the EIA effect was induced via the coupling between an electric dipole (bright mode) and an electric quadrupole (dark mode) instead of the coupling between electric and magnetic modes. However, the maximum absorption of EIA-like resonance was only about 49 % when $S' = 60$ nm. The result indicated that the slot in metal film played a decisive role for the nearly perfect absorption. A qualitative explanation was that the metal film with slots was better in blocking the transmission than the long metal bar. As shown in Fig. 4.5a, the transmission was very small for all the displacements, while the enhancement of EIA effect by increasing the displacement *S* resulted in a suppression of reflection, consequently leading to the nearly perfect absorption.

To better understand the EIA behavior, the model of TCO was calculated to fit the simulation results. The bright and the dark resonators were considered as oscillator 1 and 2, respectively. Oscillator 1 was driven by an applied field

$E(t) = E_0\exp(-i\omega t)$, while oscillator 2 was only coupled to oscillator 1. The coupling coefficient, which was a complex quantity, was equal to be $\kappa\exp(i\varphi)$, where φ was the phase shift between two oscillators due to the coupling distance in propagation direction. Therefore, the coupled differential equations were expressed as

$$\ddot{x}_1(t) + \gamma_1\dot{x}_1(t) + \omega_1^2 x_1(t) + \kappa\exp(i\varphi)x_2(t) = gE(t),$$
$$\ddot{x}_2(t) + \gamma_2\dot{x}_2(t) + \omega_2^2 x_2(t) + \kappa\exp(i\varphi)x_1(t) = 0,$$

where ω_1, ω_2, γ_1, and γ_2 ($\gamma_2 \ll \gamma_1$) were the resonant frequencies and the damping rates of oscillator 1 and 2, respectively. g indicated the strength of the coupling of oscillator 1 to the incident field. By solving the above equations under the condition of steady state, the absorption of EIA was obtained as follows.

$$A(\omega) = \frac{ig^2\omega\left(\omega^2 + i\omega\gamma_2 - \omega_2^2\right)}{\left(\omega^2 + i\omega\gamma_1 - \omega_1^2\right)\left(\omega^2 + i\omega\gamma_2 - \omega_2^2\right) - \kappa^2\exp(i2\varphi)}.$$

Assuming the detuning frequencies of both resonators were zero ($\omega_1 = \omega_2 = \omega_0$), this equation could be rewritten to second-order approximation around the resonant frequency ω_0 as

$$A(\omega) \cong g^2\left(\frac{\gamma_2\omega_0^2}{\kappa^2\exp(i2\varphi) + \gamma_1\gamma_2\omega_0^2} - i\left(\omega^2 - \omega_0^2\right)\omega_0\frac{\kappa^2\exp(i2\varphi) - \gamma_2^2\omega_0^2}{\left(\kappa^2\exp(i2\varphi) + \gamma_1\gamma_2\omega_0^2\right)^2}\right).$$

From the obtained absorption expression, it was seen that the interference term of $\exp(i2\varphi)$ was an important factor to achieve the absorption. If the interference term was positive ($0 \leq \varphi \leq \pi/4$), the absorption was smaller than that the case of no coupling ($\kappa = 0$). This corresponded to the well-known EIT effect. In contrast, if the interference term was negative ($\pi/4 \leq \varphi \leq \pi/2$), the absorption was enhanced, because of the EIA effect. Therefore, the coupling phase φ defined the spectral response exhibiting EIT or EIA behavior.

The calculated absorption spectra, which were fitted from the simulated absorption spectra by using the TCO model, nearly coincided with the simulation as shown in Fig. 4.5b. The extracted parameters from the fitted spectra were plotted as function of the displacement S in Fig. 4.6. The coupling amplitude κ was larger when the asymmetry was increased, indicating a stronger near-field coupling. However, when $S > 70$ nm, κ started to be decreased, leading to the reduction of absorption. The phase shift φ nearly remained owing to the fixed coupling distance D. The value of phase shift $\varphi \approx 0.5$, suggesting that the EIA effect was most dominant for the chosen coupling distance D. The damping rate γ_2 was smaller than γ_1, which was also consistent with the behavior of dark and bright modes in Fig. 4.2. These calculated results based on the TCO model further confirmed the aforementioned evaluation on the simulated results.

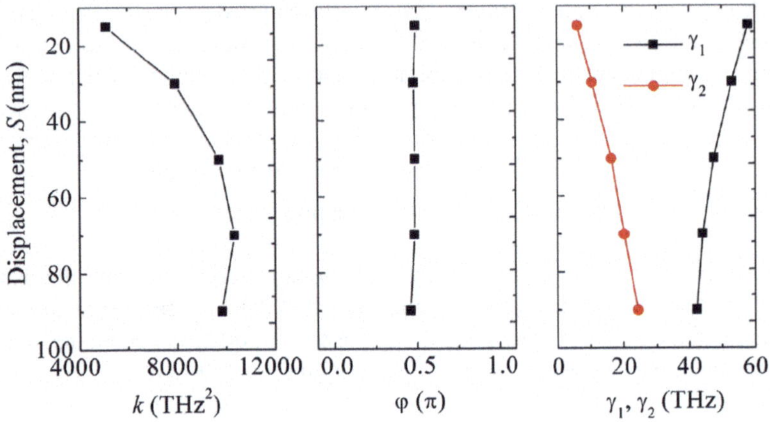

Fig. 4.6 Extracted parameters from the TCO-model fit in Fig. 4.5b as a function of displacement S

4.3 Broadband Perfect Absorbers Based on Asymmetry Resonators

A common approach to achieve broadband MMPA using the traditional sandwich structure of metal resonator/dielectric spacer/metal ground plane is designing a super cell consisting of resonators with different geometrical dimensions. If the resonances induced by these resonators are close enough in frequency, a broadband absorption is formed. Recently, through being inspired by the EIT effect in asymmetric MM, another approach to obtain a broadband absorption based on asymmetric resonators has been performed [30–33]. Figure 4.7 illustrates the unit-cell structure and the fabricated sample of a GHz MM absorber using two resonators, an inner square (IS) and an outer square ring (OSR). The detailed geometrical parameters and the materials of this MM absorber can be found in [30]. Firstly, for the symmetric structure, two absorption peaks was obtained at 4.6 and 10.6 GHz with absorption of more than 90 and 70 %, respectively, as shown in Fig. 4.8a. In Fig. 4.8b, the induce surface currents at these absorption frequencies showed that the peak at lower frequency was mostly contributed by the OSR while the IS was mainly responsible for the peak at higher frequency. Furthermore, these mainly-distributed surface currents were anti-parallel with the surface currents on the back continuous plane. Therefore, magnetic resonances were the nature of two observed absorption peaks.

For the purpose of creating a broadband absorption, an asymmetric MMPA was proposed by changing the position of IS, as shown in Fig. 4.9a, b. From the center position, the IS was moved away from the left bottom corner and toward the right top corner of the OSR. Figure 4.9c, d presented the evolution of absorption according to the displacement d of the IS. The lower-frequency absorption peak was nearly unchanged while the higher-frequency absorption peak was strongly

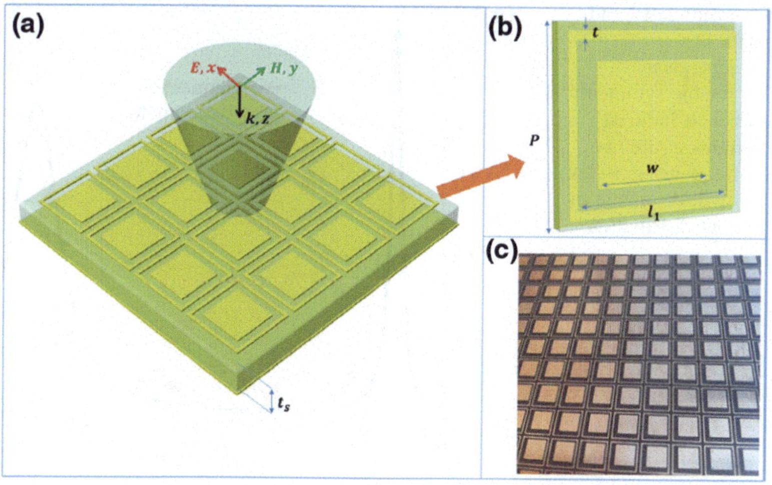

Fig. 4.7 **a** Schematic of the MM absorber and polarization configuration of the incident EM wave, **b** Single unit cell of the designed MM absorber and **c** real sample [30]

affected when d was changed from 0 to 0.8 mm. The higher-frequency absorption peak was significantly enhanced in amplitude and broader in bandwidth when $d = 0.4$ mm. The absorption was more than 90 % with a FWHM of 10 %, which was broader than that of the symmetric structure with FWHM of 5 %. Then, this peak started to be separated when d reached 0.6 mm and, finally, split into two peaks at 10.2 and 10.8 GHz when $d = 0.8$ mm. Both absorptions of these two new peaks were more than 90 %. It is noteworthy that, by finely controlling the geometrical parameters, the frequencies of two induced peaks can be very close to each other and, at the same time, the impedance of MMPA can be perfectly matched to that of the air at these frequencies. Therefore, it is possible to realize a clear broadband perfect absorption using asymmetric structure.

To clarify the nature of the resonances in the achieved triple-band absorption, the **E**-field distribution was simulated in the symmetric case ($d = 0$) and the asymmetric case ($d = 0.8$ mm). For $d = 0$ (Fig. 4.10a), the **E**-field distributions on the front layer reaffirmed that the OSR and the IS caused the low- and the high-frequency absorption peaks, respectively, and they individually behaved as dipoles (the coupling between the front layer and the back continuous plane is not discussed here). For d = 0.8 mm (Fig. 4.10b), the **E**-field at the low-frequency peak was nearly unchanged, but it moved more or less along the ring within the width t of OSR. However, the **E**-field distributions of two new separated peaks were different from that of the original high-frequency peak. The initial vertical dipole mode on the IS was degenerated into two diagonal dipole modes. Furthermore, two new quadrupole modes were excited on the OSR, which was not strongly coupled to **E**-field at high-frequency peak in the symmetric case.

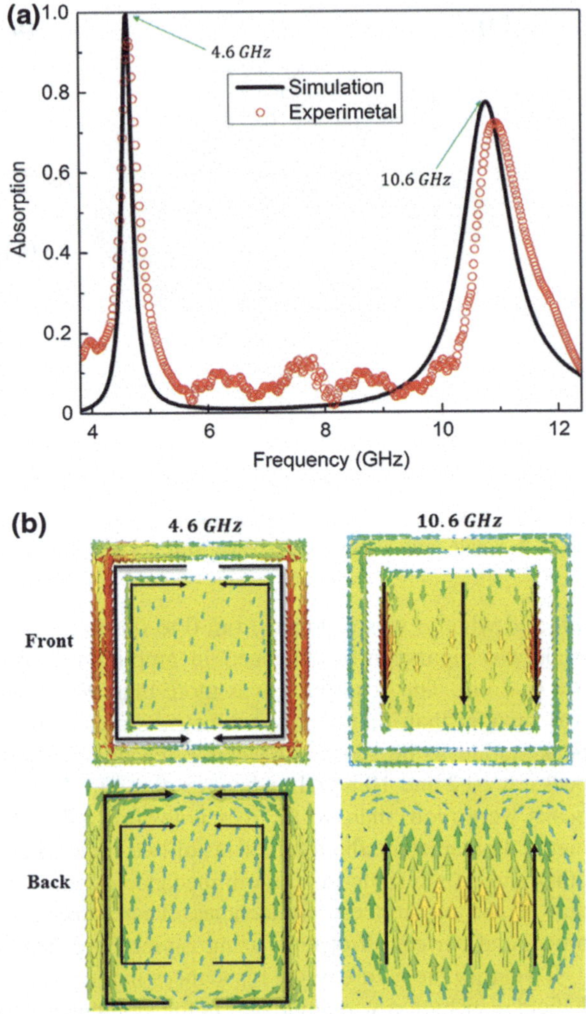

Fig. 4.8 **a** Simulated and experimental frequency-dependent absorption for the symmetric MM absorber. **b** Induced surface currents at the two resonance frequencies. *Black arrows* show directions in the surface-current distribution [30]

The **E**-field distributions at 10.2 and 10.8 GHz, and the splitting of the absorption peak can be further understood using the classical analogy of EIT, as shown in Fig. 4.11. At 10.6 GHz, the IS was strongly coupled to the incident field, giving rise to the resonance peak. This excitation corresponded to the dipole-allowed transition, $|0\rangle \rightarrow |1\rangle$ and the IS was called bright element. On the other hand, the OSR cannot be excited at the same frequency or dipole-forbidden transition, $|0\rangle \nrightarrow |2\rangle$, and it became the dark element. Therefore, for the symmetric case ($d = 0$), only single peak was observed at 10.6 GHz. Nevertheless, by breaking structural symmetry ($d = 0.8$ mm), a dark mode was excited owing to the near-field coupling between bright element and dark one, $|1\rangle \leftrightarrow |2\rangle$, resulting in a quadrupole resonance on the OSR. Finally, the interaction between the dipole resonance (bright

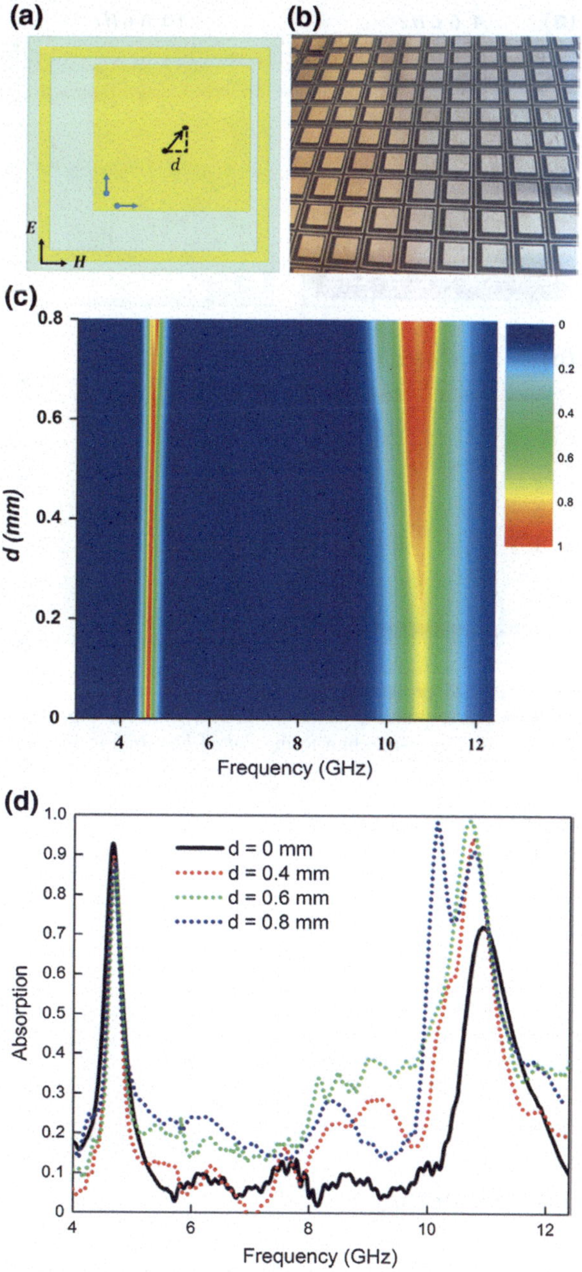

Fig. 4.9 **a** Design of the unit cell for asymmetric MMPA. **b** Photo of the fabricated sample. **c** and **d** Simulated and experimental frequency-dependent absorption for the asymmetric MMPA [30]

mode) and the quadrupole resonance (dark mode) led to the splitting of the initial resonance peak into two resonances at 10.2 and 10.8 GHz. This EIT scheme is consistent with the aforementioned **E**-field distribution in Fig. 4.10.

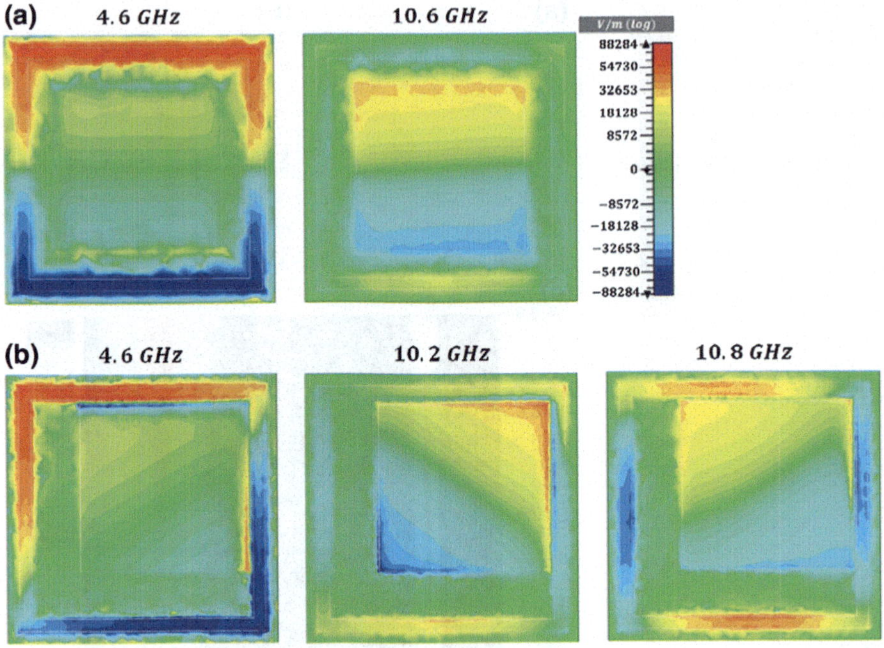

Fig. 4.10 EM properties at each resonance: distributions of induced field $\mathbf{E}_z$ of **a** symmetric structure (corresponding to the *solid-black line* in Fig. 4.9d) and **b** asymmetric structure with $d = 0.8$ mm (corresponding to the *dotted-blue line* in Fig. 4.9d) [30]

Fig. 4.11 Three-level system of EIT [30]

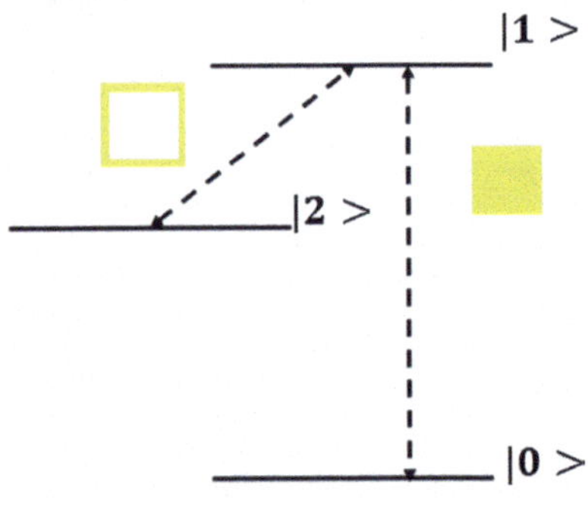

4.4 Summary

We presented some featured studies on MMPAs based on EIT and asymmetric structures. By considering the EIA effect and minimizing the transmission signal, ultra-narrow band perfect absorption was achieved, which could reveal great enhancement of EM field. On the other hand, multi-band and broadband MMPAs were realized by using asymmetric structure. Due to the symmetry breaking, new resonance modes were excited, leading to the splitting of the original absorption peak. The investigated MMPAs can be useful for potential applications in imaging, sensing, filtering, and slow-light devices.

References

1. S.E. Harris, Phys. Today **50**, 36 (1997)
2. M. Bajcsy, A.S. Zibrov, M.D. Lukin, Nature **426**, 638 (2003)
3. W. Harshawardhan, G.S. Agarwal, Phys. Rev. A **53**, 1812 (1996)
4. S. Zhang, D.A. Genov, Y. Wang, M. Liu, X. Zhang, Phys. Rev. Lett. **101**, 047401 (2008)
5. N. Papasimakis, V.A. Fedotov, N.I. Zheludev, S.L. Prosvirnin, Phys. Rev. Lett. **101**(4), 253903 (2008)
6. N. Liu, L. Langguth, T. Weiss, J. Kastel, M. Fleischhauer, T. Pfau, H. Giessen, Nature Mater. **8**, 758 (2009)
7. N. Papasimakis, Y.H. Fu, V.A. Fedotov, S.L. Prosvirnin, D.P. Tsai, N.I. Zheludev, Appl. Phys. Lett. **94**(3), 211902 (2009)
8. S.Y. Chiam, R. Singh, C. Rockstuhl, F. Lederer, W.L. Zhang, A.A. Bettiol, Phys. Rev. B **80** (4), 153103 (2009)
9. L. Zhang, P. Tassin, T. Koschny, C. Kurter, S.M. Anlage, C.M. Soukoulis, Appl. Phys. Lett. **97**(3), 241904 (2010)
10. J.J. Zhang, S.S. Xiao, C. Jeppesen, A. Kristensen, N.A. Mortensen, Opt. Express. **18**, 17187 (2010)
11. R. Singh, I.A.I. Al-Naib, Y.P. Yang, D.R. Chowdhury, W. Cao, C. Rockstuhl, T. Ozaki, R. Morandotti, W.L. Zhang, Appl. Phys. Lett. **99**, 201107 (2011)
12. J. Gu, R. Singh, X. Liu, X. Zhang, Y. Ma, S. Zhang, S.A. Maier, Z. Tian, A.K. Azad, H.-T. Chen, A.J. Taylor, J. Han, W. Zhang, Nat. Commun. **3**, 1151 (2012)
13. W. Cao, R. Singh, C.H. Zhang, J.G. Han, M. Tonouchi, W.L. Zhang, Appl. Phys. Lett. **103**(5), 101106 (2013)
14. S.H. Mousavi, A.B. Khanikaev, J. Allen, M. Allen, G. Shvets, Phys. Rev. Lett. **112**(6), 117402 (2014)
15. Y.M. Yang, Kravchenko, D.P. Briggs, J. Valentine, Nat. Commun. **5**(7), 5753 (2014)
16. N. Liu, T. Weiss, M. Mesch, L. Langguth, U. Eigenthaler, M. Hirscher, C. Sonnichsen, H. Giessen, Nano Lett. **10**, 1103 (2010)
17. F.Y. Meng, Q. Wu, D. Erni, K. Wu, J.C. Lee, IEEE Trans, Microw. Theory Tech. **60**, 3013 (2012)
18. B. Gallinet, O.J.F. Martin, ACS Nano **7**, 6978 (2013)
19. C.H. Wu, A.B. Khanikaev, G. Shvets, Phys. Rev. Lett. **106**(4), 107403 (2011)
20. K.L. Tsakmakidis, A.D. Boardman, O. Hess, Nature **450**, 397 (2007)
21. V.A. Fedotov, M. Rose, S.L. Prosvirnin, N. Papasimakis, N.I. Zheludev, Phy. Rev. Lett. **99**, 147401 (2007)
22. R. Singh, C. Rockstuhl, F. Lederer, W.L. Zhang, Phys. Rev. B **79**, 085111 (2009)
23. D.R. Chowdhury, R. Singh, A.J. Taylor, H.T. Chen, A.K. Azad, Appl. Phys. Lett. **102**(5), 011122 (2013)
24. P. Tassin, L. Zhang, R.K. Zhao, A. Jain, T. Koschny, C.M. Soukoulis, Phys. Rev. Lett. **109**(5), 187401 (2012)
25. D.R. Chowdhury, X. Su, Y. Zeng, X. Chen, A.J. Taylor, A. Azad, Opt. Express. **22**, 19401 (2014)
26. F. Hao, P. Nordlander, Y. Sonnefraud, P. Van Dorpe, S.A. Maier, ACS Nano **3**, 643 (2009)
27. N.E.J. Omaghali, V. Tkachenko, A. Andreone, G. Abbate, Sensors **14**, 272 (2014)
28. R. Taubert, M. Hentschel, J. Kastel, H. Giessen, Nano Lett. **12**, 1367 (2012)

29. J. He, P. Ding, J. Wang, C. Fan, E. Liang, Opt. Express. **23**, 6083 (2015)
30. N.V. Dung, B.S. Tung, B.X. Khuyen, Y.J. Yoo, Y.J. Kim, J.Y. Rhee, V.D. Lam, Y.P. Lee, J. Phys. D Appl. Phys. **48**, 375103 (2015). doi:10.1088/0022-3727/48/37/375103
31. B.S. Tung, B.X. Khuyen, N.V. Dung, V.D. Lam, Y.H. Kim, H. Cheong, Y.P. Lee, Opt. Commun. **356**, 362 (2015)
32. Y. Wen, W. Ma, J. Bailey, G. Matmon, X. Yu, IEEE Trans. THz Sci. Technol. **5**, 406 (2015)
33. B.-X. Wang, L.-L. Wang, G.-Z. Wang, W.-Q. Huang, X. Zhai, X.-F. Li, Opt. Commun. **325**, 78 (2014)

Chapter 5
Broadband and Tunable MMPA

Abstract Metamaterials (MMs) are artificial materials fabricated to have electromagnetic (EM) properties, which do not exist in nature. Among MMs, MM perfect absorbers (MMPAs) are promising candidates for the practical and rather immediate applications of MMs. In general, MMPA is composed of three layers. The first layer is periodically-arranged metallic patterns, whose structure and geometrical parameters should be carefully adjusted to fulfill the impedance-matching condition with the ambient, suppressing the reflection of incident EM waves. The second layer is a dielectric layer, which allows a space for the EM waves to be dissipated, and sometimes plays a role of resonance cavity to prolong the time taken by the EM waves inside the second layer. The third layer is a continuous metallic plate, blocking the remnant transmission. The properties of general MMPA are the absorption at specific frequency, the narrow absorption band and so on. Therefore, recently many researchers on MMPAs have focused on multi-band, broadband and tunable absorption. In this chapter, various researches so far about multi-band, broadband and tunable MMPAs are presented and reviewed.

Since perfect-absorption metamaterial (MM) was firstly reported by Landy et al. [1], MM absorbers have been suggested and demonstrated from microwave to optical frequencies utilizing various structures. Because the MM perfect absorber (MMPA) with a single-band is obviously inapplicable in some areas, the broadband absorption should be developed. However, the sensitive conditions, which are affiliated with optimization of the parameters of designed structure, make us difficult to satisfy the broadband perfect absorption. First of all, many researchers have worked for multi-band absorption to achieve eventually broadband absorption.

In 2009, Wen et al. [2] theoretically and experimentally reported a dual-band MM absorber in the THz range, which was composed of an electric split-ring resonator (SRR) layer and a metal-plate layer separated by a dielectric polyimide layer. They obtained that the absorption of two peaks was over 99.99 % at 0.50 and 0.94 THz for simulation, and there are two peaks of nearly perfect absorption at 0.45 and 0.92 THz for experiment (see Fig. 5.1). In the low-frequency absorption, the power loss mostly appears in the outer metallic rings and the neighboring

Y.P. Lee et al., *Metamaterials for Perfect Absorption*,
Springer Series in Materials Science 236, DOI 10.1007/978-981-10-0105-5_5

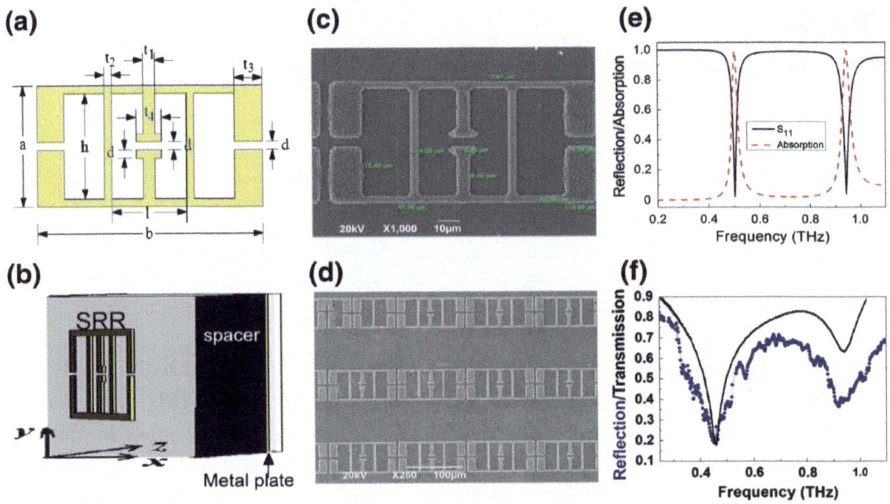

Fig. 5.1 Dual-band THz MMPA. **a** Designed electric SRR with $a = 60$, $b = 120$, $d = 4$, $h = 55$, $l = 28.6$, $t_1 = 6$, $t_2 = 4$, $t_3 = 14$, and $t_4 = 13$ μm. **b** Perspective view of the designed absorber. **c** Unit cell of the experimentally-realized absorber with electric SRR size of $a = 61.8$, $b = 119.6$, $d = 4.4$, $h = 54.6$, $l = 28.2$, $t_1 = 6.4$, $t_2 = 4.2$, $t_3 = 15.4$, and $t_4 \approx 13$ μm. **d** Photo of a portion of the fabricated absorber. The size of absorber unit cell is 140 × 124 μm^2. **e** Simulated reflection (*solid line*) and absorption (*dotted line*) curves of the absorber. **f** Measured reflection curve of the absorber (*symbols*) and simulated transmission curve of the electric SRR (*smoothened line*) [2]

polyimide slab, with the strongest absorption at the fringe of the two outer split gaps (shown in Fig. 5.2). On the other hand, for the high-frequency absorption, most of the power loss take place around the inner SRR with only a small portion occurring at the outer split gap.

Zhang et al. [3] proposed numerically and experimentally the polarization-independent dual-band MMPA in near-infrared regime by using elliptical nanodisks. The dual-band absorption appeared utilizing elliptical nanodisks, which consists of two resonance peaks with respect to their major and minor axes. For the longer wavelength absorption peak, for both polarizations, the measured results revealed a resonance wavelength at λ_1 = 1703 nm, identical to the simulation anticipating 1679 nm, and the measured absorption (93.2 %) was slightly decreased with respect to the simulation of 99 % (not shown here). For the shorter absorption peak, the difference between simulation and experiment was 27 nm for both polarizations (see Fig. 5.3). The resonance wavelength shifted from the simulation of 1247 nm to the measured λ_2 = 1274 nm, and the absorption was reduced to be 89.3 %.

Liu et al. [4] investigated selective thermal emitters in the mid-infrared range based on MMPA (Fig. 5.4). The metallic ground plane was fabricated to be thick enough to avoid light transmission, and therefore a narrow band absorber was obtained with high absorption. The design might be simply tuned to be for longer or shorter wavelength owing to the fact that the resonance frequency of the cross

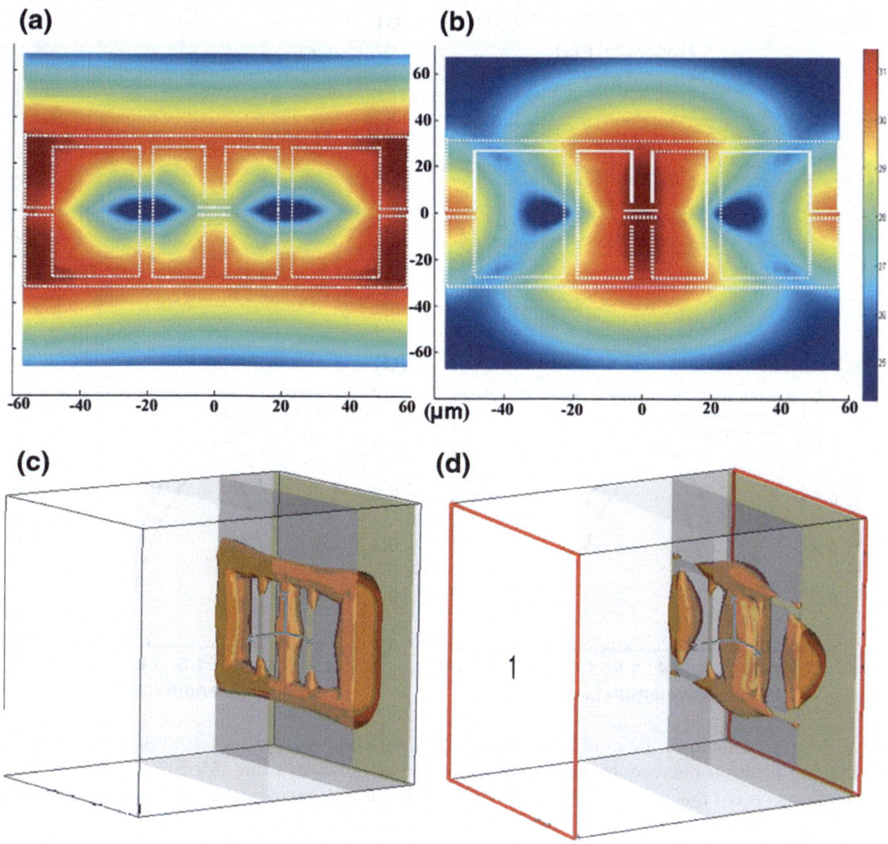

Fig. 5.2 Distribution of the power-loss (absorption) of the THz wave in the absorber. Bird-view for **a** low-frequency and **b** high-frequency absorption, and 3-dimensional distribution for **c** low-frequency and **d** high-frequency absorption [2]

resonator was relative to its length. Two different MMPAs were combined to form the bipartite checker-board unit cell. Figure 5.4e, f show that the single-band absorber presents an absorption peak at 5.8 μm with 97 % absorption and two absorption bands appear by the dual-band absorber at 6.18 and 8.32 μm with 80 and 93.5 % absorption, respectively. The emittance of the surface over a broad band was engineered, as shown in Fig. 5.5. The emittance to follow the EQE of GaSb was designed in a wavelength-dependent fashion.

A novel single-patterned-layer dual-band MMPA with a simple unit-cell structure has been suggested for mid-IR by Jiang et al. [5]. The dual-band MMPA was composed of three layers: a solid ground plane at the bottom, a thin dielectric layer, and a doubly-periodic array of electrically-isolated nanoresonators at the top. Figure 5.6a show a detailed structure illustration. The top Au screen consist of a doubly-periodic array of orthogonal stub-loaded H-shaped nanoresonators realized by using a genetic algorithm, which have a central connecting bar. The

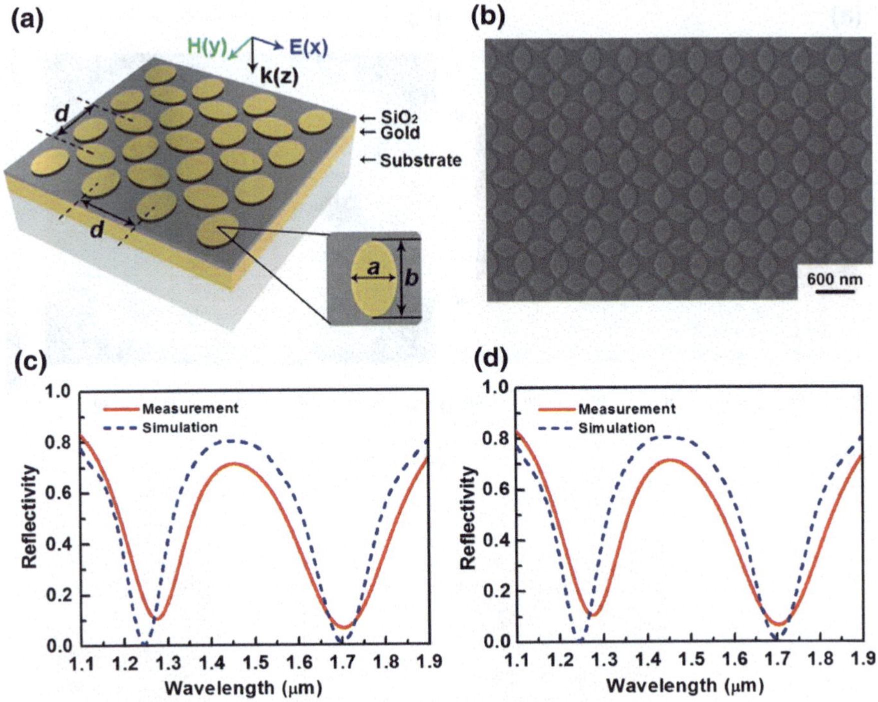

Fig. 5.3 **a** Schematic of the dual-band MMPA structure and incident polarization configuration. **b** Scanning-electron-microscopy (SEM) image of the designed structure. Measured and simulated reflection spectra of the dual-band absorber for **c** TE and **d** TM polarization at normal incidence [3]

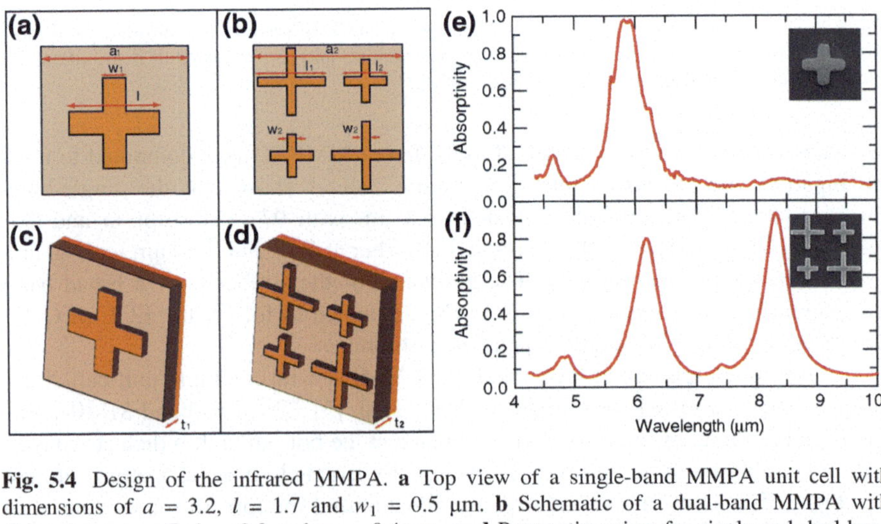

Fig. 5.4 Design of the infrared MMPA. **a** Top view of a single-band MMPA unit cell with dimensions of $a = 3.2$, $l = 1.7$ and $w_1 = 0.5$ μm. **b** Schematic of a dual-band MMPA with dimensions: $a_2 = 7$, $l_1 = 3.2$ and $w_2 = 0.4$ μm. **c**, **d** Perspective view for single and dual-band MMPA. Thickness of dielectric spacer was $t_1 = 0.2$ and $t_2 = 0.3$ μm. **e** Experimental absorption of the single-band MMPA. **f** Experimental absorption of the dual-band MMPA. *Insets* display the SEM images of unit cell for the fabricated single- and dual-band absorbers [4]

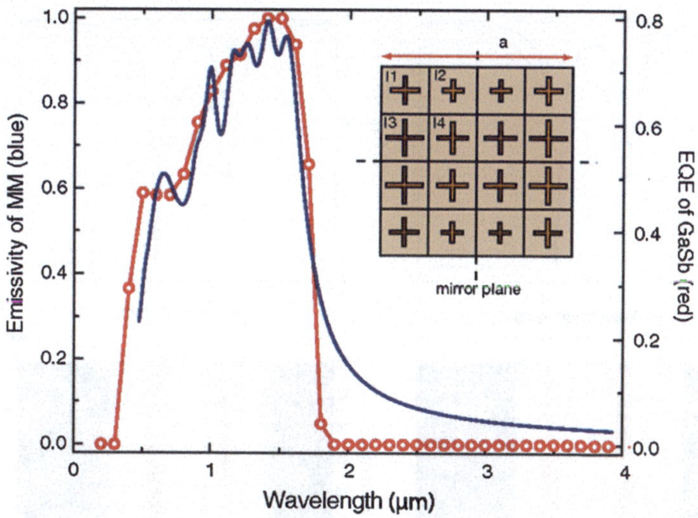

Fig. 5.5 Designed MM emissivity (*blue solid curve*) and external quantum efficiency (EQE) (*red circle curve*) of GaSb. *Inset* shows a schematic of the MM design with 16 sub-lattices. Dimensions of the design were $a = 1080$, $l_1 = 173$, $l_2 = 127$, $l_3 = 221.5$ and $l_4 = 190$ nm. Each cross sub-lattice had a line width of 20 nm and thickness of 50 nm. The thickness of dielectric spacer was 78.8 nm and of ground plane was 200 nm. The unit cell also possessed two mirror planes, one *horizontal* and *one vertical*, as indicated [4]

directions of the incident TE and TM polarized fields regarding the MMPA are illustrated in Fig. 5.6a. This design fulfills the nanofabrication design restrictions assigned during design optimization, which confirms that the specific structure can be performed experimentally to minimize differences that would reduce the resonant EM properties compared with theory. Figure 5.7a shows that the simulated

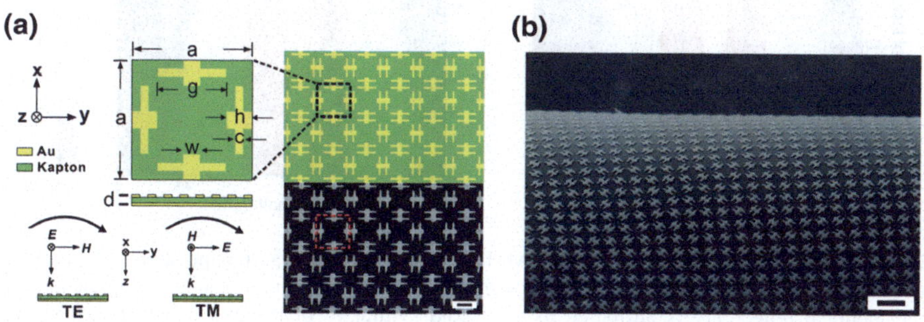

Fig. 5.6 Diagram and field-emission SEM (FESEM) image of the fabricated dual-band mid-IR MMPA coating. **a** *Top* Doubly-periodic array of H-shaped nanoresonators showing magnified view of one unit cell with dimensions $a = 1475$, $h = 315$, $w = 210$, $g = 840$, $c = 105$ and $d = 200$ nm (*top* and *bottom* Au: 50 nm and Kapton: 100 nm). *Bottom right* FESEM image of a portion of the fabricated MMPA. Scale = 600 nm. *Bottom left* Orientation of the incident fields with respect to the MMPA. **b** Low-magnification FESEM image of the freestanding fabricated conformal MMPA coating showing its mechanical flexibility. Scale = 1800 nm [5]

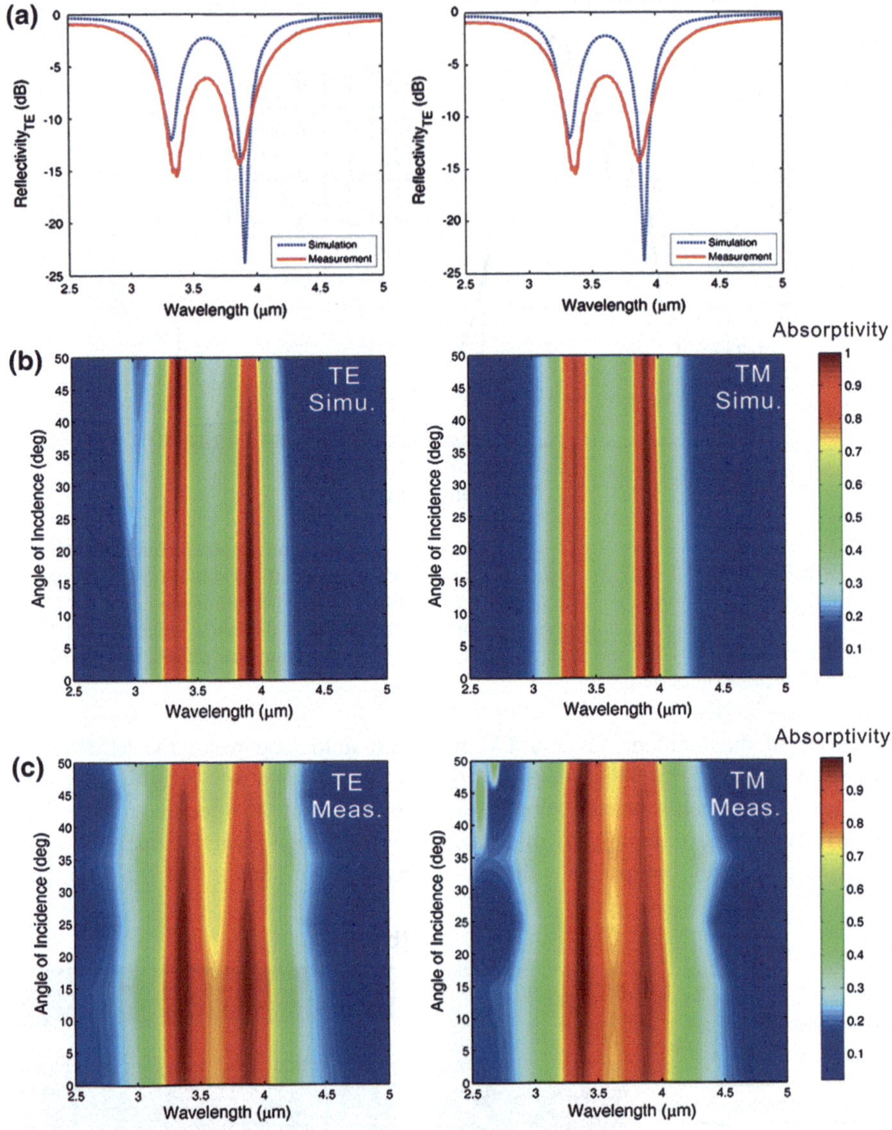

Fig. 5.7 Angular peak dispersion of the MMPA absorption spectra. **a** Simulated and measured reflectivity of the MMPA coating under TE (*left*) and TM (*right*) illumination at normal incidence. The identical reflectivity is attributed to the 8-fold symmetry of the unit cell that enables the response of the structure to be independent of the azimuthal angle of the incoming light. **b** Contour plot of the simulated absorptivity as a function of wavelength and angle of incidence under TE (*left*) and TM (*right*) illumination. The two vertical *red strips* clearly show two strong absorption bands nearly independent of the incident angle. **c** Contour plot of the measured absorption as a function of wavelength and angle of incidence under TE (*left*) and TM (*right*) illumination. The two angularly-independent absorption bands agree well with the theoretical predictions. Near-unity absorption was achieved in both bands experimentally, confirming a successful realization of the designed structure [5]

reflection of the dual-band MMPA for normal incident wave for the TE and the TM polarizations. Transmission is zero because of the Au ground plane. Two strong absorption bands distinctly appear at the target wavelengths of 3.3 and 3.9 μm. Both bands possess −10 dB bandwidth of ∼0.1 μm with a maximum absorption of 94.7 % or −0.24 dB (reflection of 6.3 % and −12 dB) at 3.3 μm and of 99.6 % or −0.02 dB (reflection of 0.4 % and −24 dB) at 3.9 μm. Figure 5.7b presents that the simulated absorption spectra for the TE and the TM polarizations. The two vertical red strips manifest that the two absorption peaks are maintained, centered at 3.3 and 3.9 μm over a wide range of incidence angles for both TE and TM polarizations. The absorption in both bands comes to be >91 % even by changing the incident angle up to ±50°, due to the efficient excitation of both electric and magnetic resonances.

The triple-band MMPA in the THz range was theoretically and experimentally investigated by Shen et al. [6]. Al film (0.2-μm thick) was thermally evaporated on an n-type Si substrate (640-μm thick), then a polyimide layer (7.5-μm thick) was put on the Al film in order to form the spacer layer. The Al ring patterns (0.2-μm thick), which consisted of three concentric closed rings to form a compact single particle, were engineered by ordinary optical lithography on the polyimide (as shown in Fig. 5.8a, b). In Fig. 5.8c, the absorption spectrum had three peaks at frequencies of 0.5, 1.03 and 1.71 THz with 96.4 %, 96.3 % and 96.7 %, respectively. They demonstrated the absorption mechanism, which was the reasonability of the multi-reflection theory to analyze the MMPA (see Fig. 5.9). The destructive interference of multi-reflection beams led to nearly zero reflection, and therefore the appearance of nearly unity absorption at specific frequencies.

In 2012, Xu et al. [7] reported numerically and experimentally alternative strategy to gain sufficiently smaller MM elements and a comprehensive transmission-line analysis on a triple-band absorber from a concept enhanced by a

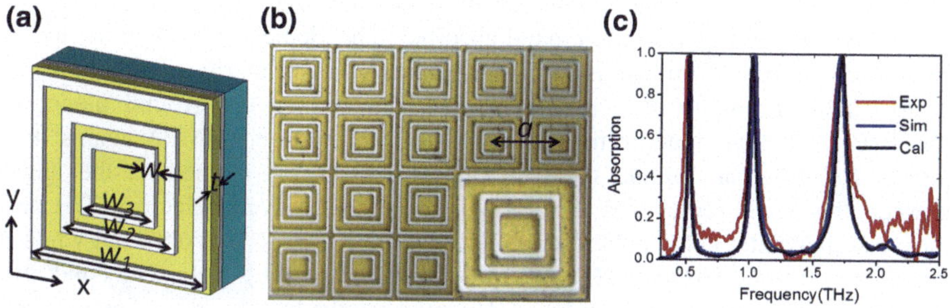

Fig. 5.8 **a** Triple-band MM absorber structure. Metallic ground plane and sub-wavelength particles (*white*) were Al and spacer (*yellow* or *dark in black and white*) between them was polyimide. The side lengths of sub-wavelength particles on the top layer were denoted as w_1, w_2 and w_3, and the line widths are all set as w. **b** Optical-microscopy image of a portion of the fabricated structure. Periods in both dimensions were the same and were denoted as a. The *inset* shows the enlarged unit cell. **c** Comparison among the experimental absorption (*red line*), full-wave simulated absorption (*blue line*), and theoretical multi-reflection absorption (*black line*) of the THz triple-band MM absorber [6]

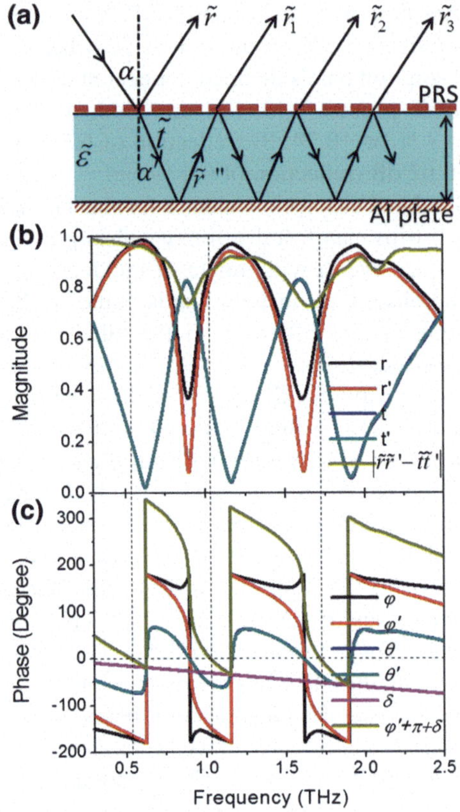

Fig. 5.9 **a** Physical model to compute the complex reflection and transmission coefficients at interfaces. **b** Simulated magnitudes, and **c** phases of the reflection and transmission coefficients for the decoupled model. Coefficients r and t (φ and θ) are the magnitudes (the phases) of reflection and transmission as the incident wave reached the air/part reflection surface (PRS)/spacer. Coefficients r' and t' (φ' and θ') are the magnitudes (the phases) of reflection and transmission of the wave reflected by the ground plane, reaching the spacer/PRS/air. δ is the phase delay caused by the round propagation in the spacer [6]

two-dimensional (2D) left-handed transmission line. Figure 5.10a–d show the schematic and the equivalent-circuit model of the proposed MM element that was employed in the design of a triple-band absorber. The element particle consisted of three layers: a front 2D transmission line element which primarily reacted with the electric field, a middle dielectric layer, and a back metallic ground plane. In Fig. 5.10e, f, the simulated reflection curves of the proposed MMPA against frequency for different oblique incident angles, which was defined as the angle between the wave vector and the normal, for both transverse-electric (TE) and transverse-magnetic (TM) polarizations. A negligible change was obtained when θ varied from 0 to 60° in both TE and TM polarizations at 2.09 GHz. The condition was slightly different at 6.53 and 10.3 GHz owing to the fact that the reflection dips change continuously and a series of additional dips were exhibited around those frequencies. The reflection under normal incident planar TE wave with different polarizations, which was defined as the angle between the E field and x axis, were plotted in Fig. 5.10g. Owing to the high degree of four-fold rotational symmetry, the intensity of reflection dips remained unchanged when φ increased from 0 to 90°.

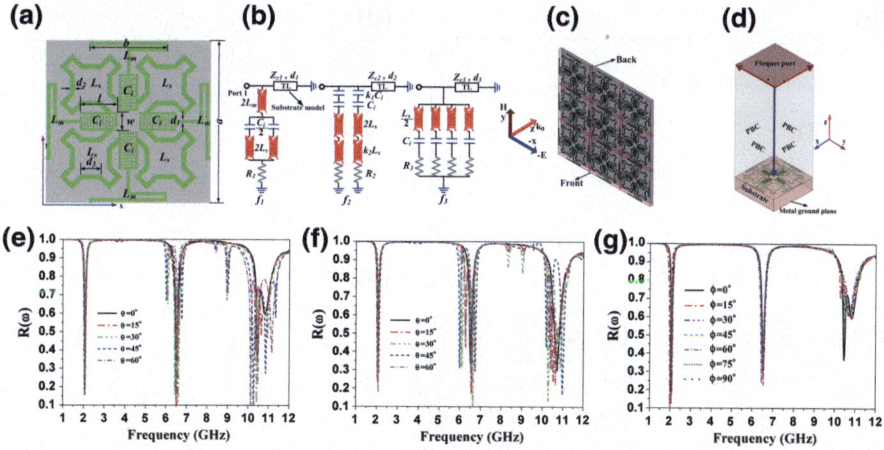

Fig. 5.10 **a** Topology and **b** equivalent-circuit models of the proposed MMPA structure at three principle resonances. **a** Top view of the MM transmission-line (TL) element. **b** Triple-circuit models corresponding to three individual resonances. **c** View of a portion of the designed MMPA and **d** simulation setup in High-frequency Structural Simulator. Geometrical parameters were $a = 10.6$, $b = 5$, $l = 2.4$, $w = 1.2$, $d_1 = 0.2$, $d_2 = 0.3$, and $d_3 = 1.32$ mm. Length of the meandered 3/2 fractal curve was $L = 6.8$ mm. Angular independence of the reflection of the proposed MMPA for **e** TE and **f** TM wave, respectively. **g** Polarization independence of the reflection of the proposed MMPA for TE wave excitation [7]

Park et al. [8] explained theoretically multi-band absorption mechanism and suggested the quadruple-band MMPA in the GHz range by using simple donut structures. The samples were designed by the conventional printed-circuit-board process with metallic patterns on a plane of incidence and a metallic plane on the other side, separated by a dielectric FR-4 layer. The simulation results revealed triple absorption peaks at 6.5, 7.0 and 7.6 GHz with absorption of 99 %, 98 % and 99 %, respectively. The corresponding experimental absorption peaks were shown at 6.51, 7.0 and 7.61 GHz with absorption of 98 %, 98 % and 98 %, respectively. It was confirmed that both simulation and experiment were in excellent coincidence (as in Fig. 5.11). By the free-space impedance matching of MMPA and the prevention of transmission, electromagnetic (EM) wave might be confined in the copper-FR-4-copper structure. The magnetic resonance, which was related with the effective permeability, could appear by the anti-parallel surface currents in two metallic layers, and the effective permittivity was formed by each metal structure. In addition, most of the power loss occurred in the upper and the lower parts of donut, and the dielectric losses took place in between the two metallic plates. They analyzed the effect of interferences among the multiple-reflection beams between two metallic layers, but found that the multiple-reflection theory was not suitable for explanation of the absorption mechanism of donut structure (shown in Fig. 5.12). Figure 5.13 displays that the optimized pattern consisted of 4 donut-type structures which had different radii. The simulation results revealed quad absorption peaks at

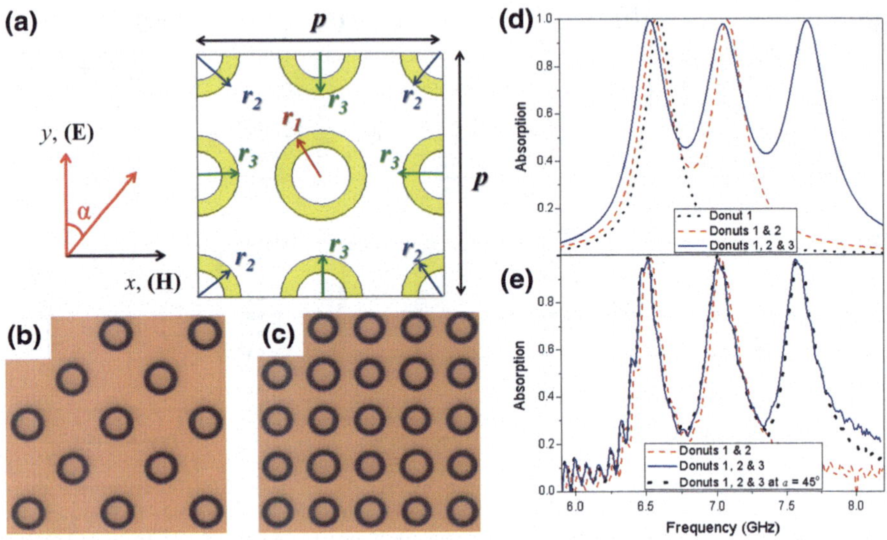

Fig. 5.11 **a** Schematic of the proposed structure of MMPA. Photos of the fabricated samples with **b** 2 types and **c** 3 types of structures. **d** Simulated and **e** measured absorption spectra for the combinations of donuts [8]

6.45, 7.4, 9.1 and 11.0 GHz with absorption of 97 %, 98 %, 98 % and 98 %, respectively. The corresponding experimental absorption peaks were revealed at 6.5, 7.4, 9.2 and 11.0 GHz with absorption of 98 %, 97 %, 98 % and 98 %, respectively.

Yoo et al. [9] investigated the polarization-independent dual-band MMPA by using only one kind of pattern. Figure 5.14a, b show the schematics for the two suggested MMPA structures: disk-type and donut-type, and the photos of fabricated structures for the disk-type and the donut-type structures were shown in Fig. 5.14c, d, respectively. Two high-absorption peaks were evident. For disk-type absorber, the low-frequency peak (absorption of 94 % at 6.22 GHz) resulted from the fundamental magnetic response and the high-frequency one (absorption of 93 % at 17.38 GHz) was induced by the third-harmonic resonance of meta-atom. For donut-type absorber, in common with disk-type, the low-frequency peak (absorption of 99.6 % at 5.10 GHz) was induced by the fundamental magnetic response and the high-frequency one (absorption of 99.8 % at 16.32 GHz) was caused by the third-harmonic resonance. The donut-type structure displayed absorption of 99.1 % at 5.06 GHz and of 99.9 % at 16.88 GHz in experiment. It was confirmed that both simulation and experiment were in good coincidence (Fig. 5.14e, f).

To achieve a broadband (not multi-band) MM absorber, many researcher have investigated MMPA for the broadband utilizing various structures. In 2010, Wakatsuchi et al. [10] numerically reported that the customized broadband absorption of EM waves having arbitrary polarization was possible by use of lossy cut–wire (CW) MM. The CW structure displayed very effective absorption

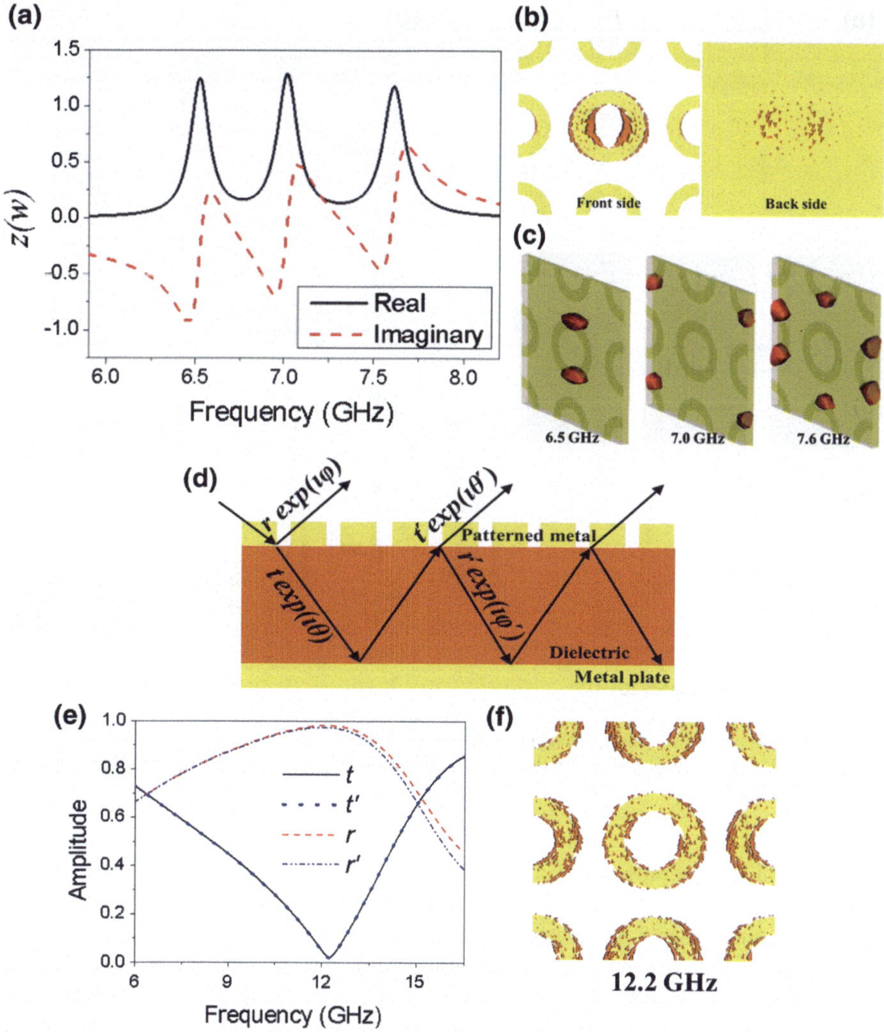

Fig. 5.12 **a** Simulated effective surface impedance of the triple-band absorption. **b** Distribution of the normalized induced current at a resonance frequency (6.5 GHz). **c** 3D distribution of the power loss density at each resonance frequency. **d** Schematic diagram for multiple-reflection theory with associated parameters. **e** Amplitude of the reflection and the transmission coefficients at the air-dielectric interference for the system of only resonator array and dielectric (decoupled model). **f** Normalized surface current distribution of the decoupled model at 12.2 GHz [8]

characteristics, and it was shown that different lengths of CWs, which were combined in one unit cell of MM absorber, made broadband absorption. The absorption property might be further enhanced by optimizing the sheet-resistance values, and use of additional pairs of CWs was employed not only to increase absorption, but also to decrease it (Fig. 5.15).

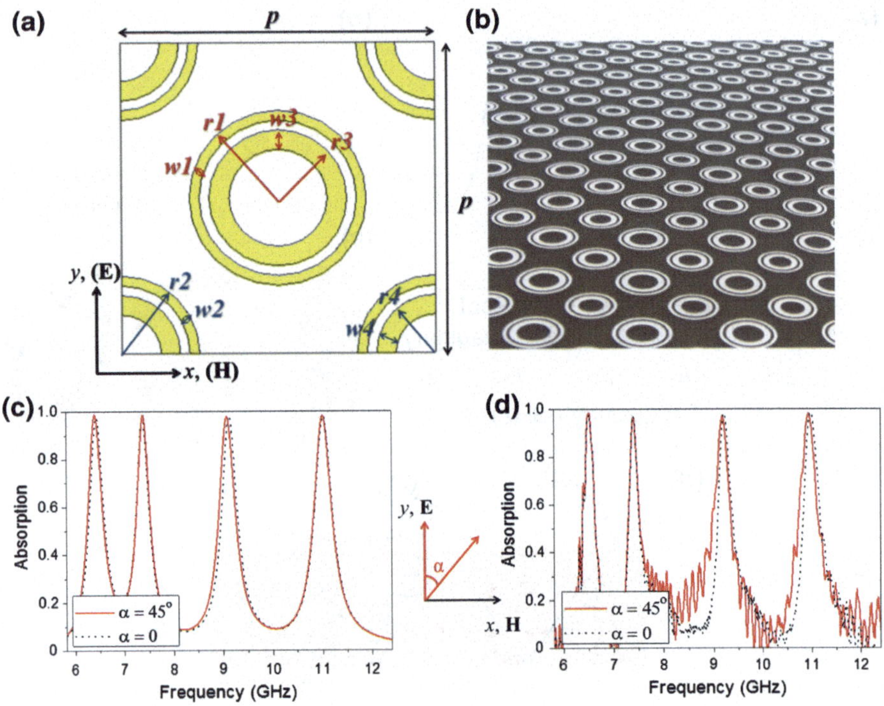

Fig. 5.13 **a** Schematic of the proposed structure of MMPA for quad peaks and **b** photo of the fabricated sample. **c** Simulated and **d** measured absorption spectra of quad-peak MMPA according to polarization [8]

Sun et al. [11] numerically and experimentally investigated a broadband MM absorber, which was composed of lossy frequency-selective surface (FSS) and a metallic ground plane, separated by a dielectric layer. Figure 5.16a, b display schematically the unit cell of the designed absorber as well as the propagation configuration of incident EM wave, and the corresponding circuit model. The design results showed that the bandwidth for the reflection below −10 dB with 3- and 4-mm thick absorber came to be 6.6–18 and 5.27–18 GHz, respectively, which were wider than that of the corresponding absorber with conventional FSS pattern. Three resonances (f_1, f_2 and f_3) occurred between FSS and dielectric substrate, which displayed that the absorber had three absorption nulls (in Fig. 5.16c). The experiments were carried out to confirm the design results. Three absorption nulls existed in a considerable frequency range, which was accorded with the design result (as in Fig. 5.17).

In 2011, Cui et al. [12] reported that an infrared broadband absorber based on an array of nanostrip antennas of several different sizes. A 100-nm thick gold film was deposited on the pure glass substrate by e-beam evaporation, and a 340-nm thick Ge film was prepared by the same method. By using e-beam lithography and lift-off technique, the gold nanostrip antennas of four different widths (labelled by W_1–W_4)

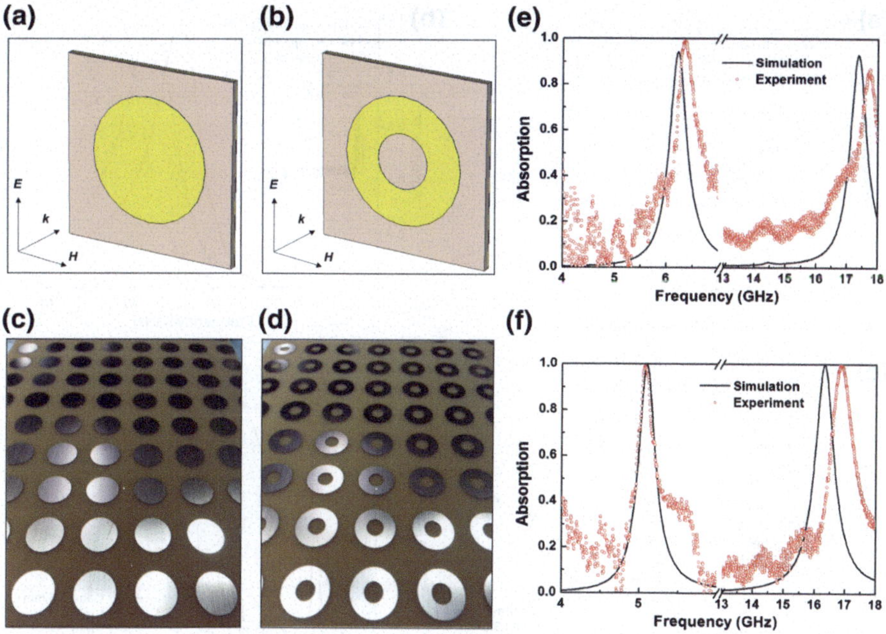

Fig. 5.14 Unit cells for the MMPAs of **a** disk-type and **b** donut-type structures. Photos of the fabricated samples with **c** disk-type and **d** donut-type structures. Simulated and measured absorption spectra for **e** disk-type and **f** donut-type structures [9]

within a consistent period were patterned at the top (see Fig. 5.18a, b). Figure 5.18c, d show that the absorbing spectrum of the multi-sized antenna absorber was very wide owing to the merge of four close peaks (I, II, III and IV) together, of which peak III was very close to the resonant peak of the single-antenna absorber. Figure 5.19a displayed the absorption spectra as tuning difference Δ from 0 to 200 nm while other parameters were fixed, with the dashed lines indicating the cases of $\Delta = 0$, 42.5, 85 and 127.5 nm. In Fig. 5.19c, both spectrum band and the absorption strength read from the calculated spectra have been reproduced in the experimental results.

Ye et al. [13] have investigated that MM could be created to have frequency-independent properties over an ultra-wide frequency range. In Fig. 5.20a, the basic sub-wavelength unit cell (Cell 1) consisted of metallic SRR and rod. In order to introduce multiple electric and magnetic resonances without increasing the size of cell, the original SRR was split into four smaller, centrosymmetric SRRs, shown as ‘Cell 2’. The outside corner of each SRR was split again and a lumped resistor was brazed to introduce independently tunable loss for each SRR, as illustrated by ‘Cell 3’. In order to clearly display the effect of dispersion control, an MMPA was made to consist of the cells (shown in Fig. 5.20b). The near-unity impedance of the MMPA layer was formed in the entire grey region (1.25–1.84 GHz), nearly similar to the impedance of air, and the reflection coefficient was

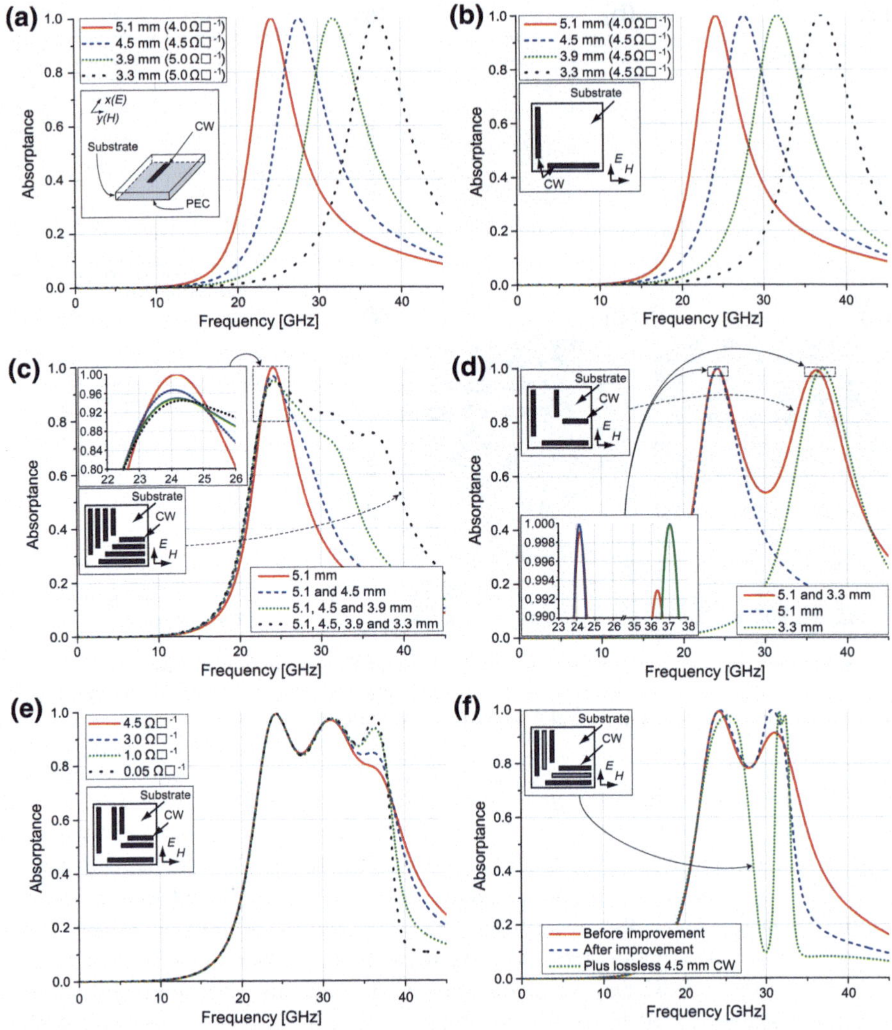

Fig. 5.15 Absorption of single-CW MM placed on a perfect electric conductor (PEC) wall. **a** Absorptions for different length of CW. The *inset* describes the simulated situation. **b** Absorption when another pair of the CW used in **a** was placed orthogonally to the first as illustrated in the *inset*. **c** Absorption when part or all of the CW pairs were combined as one MM unit. The *inset* shows the structure having all the CW pairs. **d** Use of two CWs exhibits a double absorption peak close to the individual peaks. **e** Illustrates absorption of three CW pairs (5.1, 3.9 and 3.3 mm). Modification of the resistance value used for the 3.3-mm CW led to a triple absorption peak. The resistance values used for the 3.3-mm CW are shown in the legend. The other resistance values were the same as those of **b**. **f** Absorption of two CW pairs of 5.1 and 3.9 mm was improved by use of adjusted resistance values. In addition, the use of a lossless 4.5-mm CW pair exhibits a strong reduction of the absorption around about 30 GHz [10]

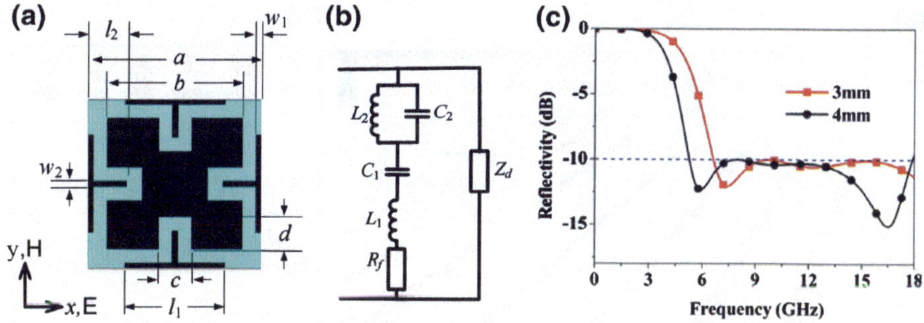

Fig. 5.16 **a** Schematic structure of the designed FSS and **b** its corresponding circuit model. **c** Optimized reflection of the designed absorber [11]

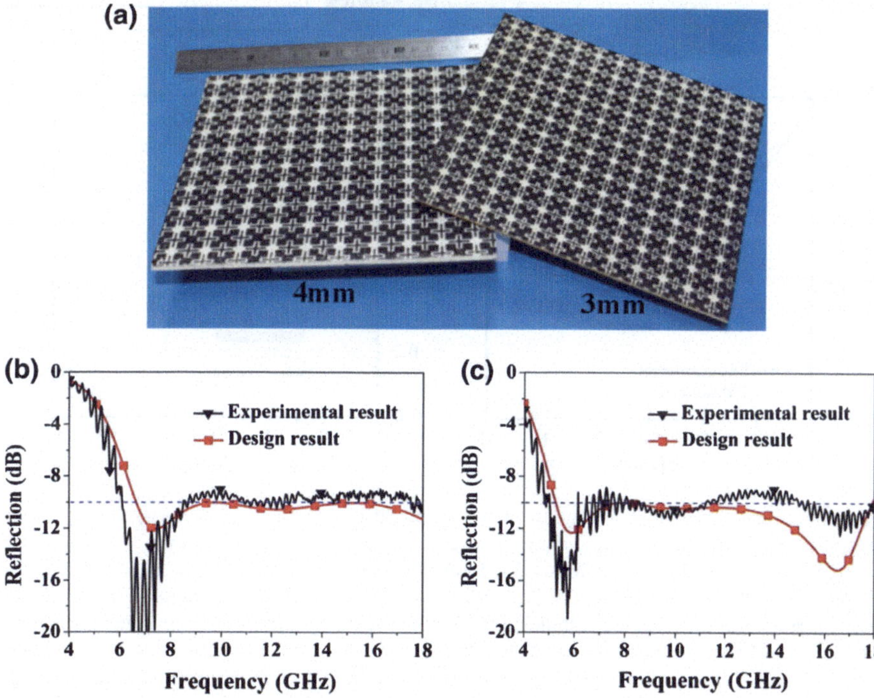

Fig. 5.17 **a** Fabricated sample of 3- and 4-mm broadband lossy FSS absorber. Comparison of the reflection of the broadband absorber between experimental (*black solid curve*) and simulated (*red square curve*) results, for **b** the absorber thickness of 3 mm and **c** of 4 mm [11]

lower than −30 dB in this region. The lumped resistances was important for the precise dispersion control to gain nearly perfect absorption (see Fig. 5.20c). By increasing the four resistances, the second magnetic resonance and the two electric resonances inside the grey region were indistinguishable owing to the increased

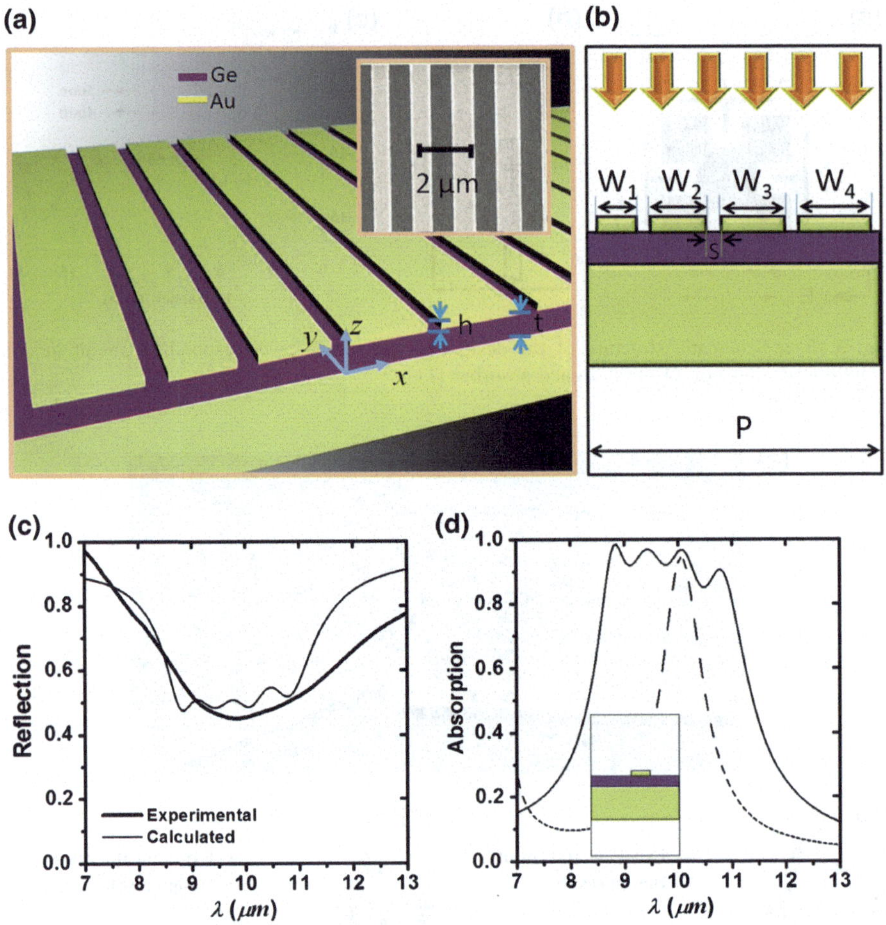

Fig. 5.18 **a** and **b** Schematic diagram of the proposed three-layered light harvesting absorber. *Inset* of **a** shows the SEM image of the fabricated sample with P = 6.06 μm, t = 340 nm, and h = 16.5 nm. Strip widths (from W_1 to W_4) form an arithmetic sequence, with W_3 = 830 nm, the difference Δ = 85 nm where $\Delta = W_{n+1} - W_n$ (n = 1, 2, 3), and the separation S = 727.5 nm. **c** Experimental (*thick curve*) and simulated (*thin curve*) reflection spectra for the sample presented in Fig. 5.28. **d** Comparison of the calculated absorption spectra between multi-size antenna (*solid curve*) and the single-sized antenna (*dashed curve*). The latter profile is shown in the *inset* [12]

dissipation, and both imaginary and real parts inside the grey region flattened out (Fig. 5.20d). In the case of incident angles below 30°, the power reflection below −20 dB still occurred over an ultrawide bandwidth of 0.59 GHz (1.22–1.81 GHz), in accordance with a relative bandwidth of 39 % (see Fig. 5.21).

Zhu et al. [14] demonstrated an ultra-broadband, polarization-insensitive, and wide-angle MM absorber at THz utilizing arrays of truncated pyramid unit structure composed of metal-dielectric multilayer composite (see Fig. 5.22a). Figure 5.22b–e

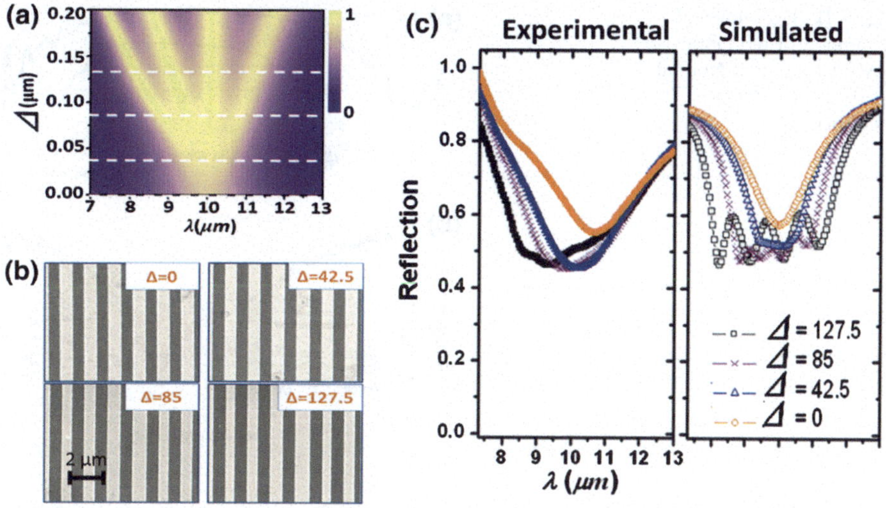

Fig. 5.19 **a** Absorption spectra when tuning the difference Δ from 0 to 200 nm with *four dashed lines* denoting the cases of $\Delta = 0$, 42.5, 85 and 127.5 nm. For the four cases, **b** the SEM images and **c** the experimental and simulated overall reflection spectra [12]

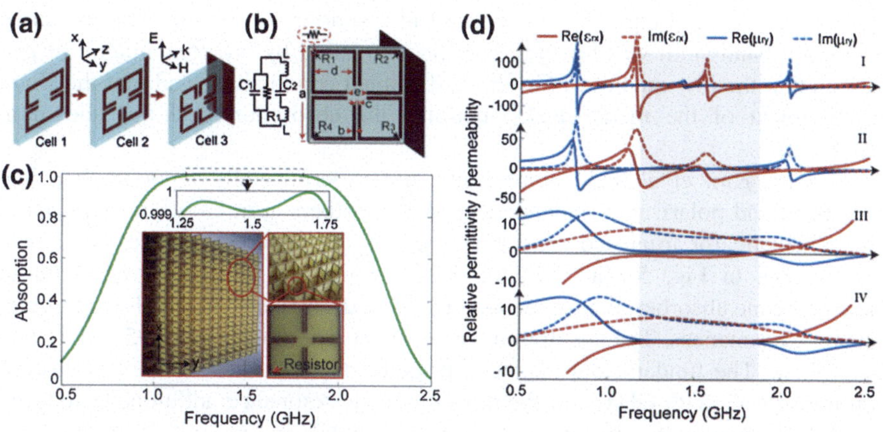

Fig. 5.20 Design of the unit cell of an ultrawide-band MMPA. **a** Introduction of multiple resonances in an SRR and rod cell. **b** Topology of the unit cell and related equivalent-circuit model. **c** Full-wave simulation of the fabricated MMPA sample. **d** Retrieved constitutive parameters for different resistances [13]

show the absorption spectra according to the incident conditions for both TE and TM polarizations. In the case of TE polarization, absorption intensity decreased by increasing the incident angle above 30°, on the other hand, the bandwidth was only slightly changes. In the case of TM polarization, the absorption intensity remained according to the incident angle even up to 70° and two narrow absorptions at

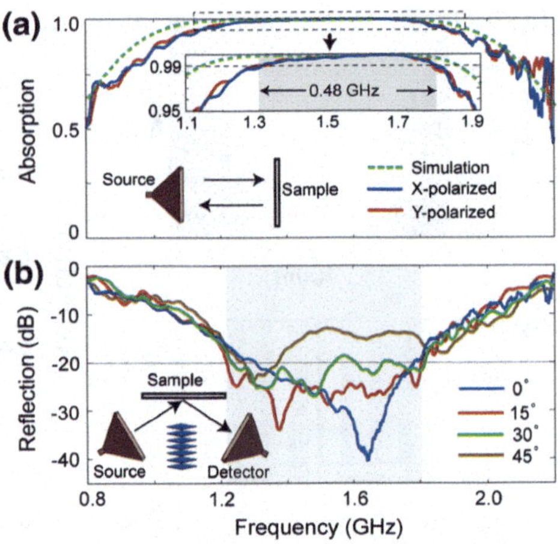

Fig. 5.21 Experimental absorption of the fabricated MMPA sample. **a** Normal incidence. **b** Oblique incidences [13]

frequencies above 1.6 THz, which might be contributed by higher-order Bloch modes excited by truncated pyramid structures with period comparable with the applied wavelengths at higher frequencies. Figure 5.22f–m present the electric and the magnetic field distributions at specific frequencies. The first-order parallel-plate waveguide mode became excited in different layers at different frequencies, and the mode with a lower frequency was excited in a shorter waveguide. The measured absorption bandwidth was broader than the simulation by the combination of two reasons that the low quality of Cr film had a small conductivity and that the slight misalignment of the metal stacks modified the absorption peak (as shown in Fig. 5.23).

In 2015, Kim et al. [15] numerically and experimentally reported that dual wide-band and polarization-independent MM absorbers at microwave frequencies. The schematic for truncated-cone structure and the picture of fabricated sample were shown in Fig. 5.24a–c. Figure 5.24d shows the absorption spectrum of truncated-cone absorber, whose meta-atoms were designed to enable the high-order magnetic resonances. The absorption was over 90 % in both 3.93–6.05 and 11.64–14.55 GHz. The fundamental magnetic response made the low-frequency absorption band (3.93–6.05 GHz) and the third-harmonic resonance additionally induced the high-frequency absorption band (11.64–14.55 GHz). Figure 5.25 displays the simulated and the measured absorption spectra for polarization of 0, 30° and 45°. The absorption was experimentally decreased more or less in 9.95–10.46 GHz because of non-zero incident angle and values of dielectric constants.

Many researcher have also studied tunable MMPAs, which are required for various applications, as well as broad-band MMPAs. A tunable hybrid MM absorber in the microwave band was realized by Wen et al. [16]. In Fig. 5.26a, b, the designed MM absorber had a similar structure to previously-proposed dual-band THz absorber [2], except that a thin VO_2 film was included between electric SRR (eSRR) and substrate. The unit cell consists of two types of split gaps with the inner resonator and the outer resonator pairs. Figure 5.26c displays that the calculated

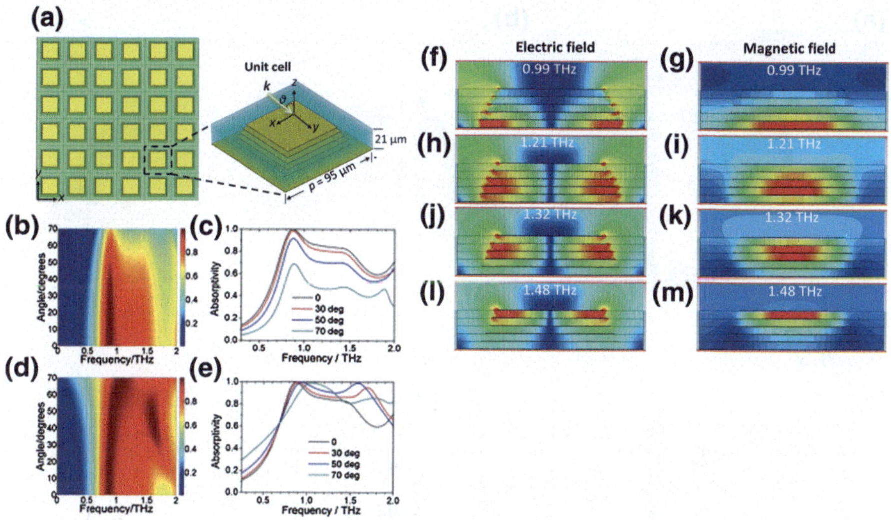

Fig. 5.22 **a** Schematic of THz MMPA design. *Left* Top view of the structure. *Right* Oblique view of the unit cell. Design includes a square array (period of 95 µm in the *x*-*y* plane) of truncated pyramid element made of five metal films stacked along the z axis and in between separated by a 4-µm-thick dielectric polymer. In this prototype, the square metallic patches from the top to the bottom had a side width of 54, 58, 62, 66 and 70 µm. The sample had a total thickness of 21 µm. A 200-nm-thick opaque metal wall was used as the background mirror. The incident wave was assumed in the *x*-*z* plane at an incident angle θ. Simulated absorption. **b** and **c** are for TE polarization, and **d** and **e** are for TM polarization. In **b** and **d**, the *x* and *y* axes represented frequency and incident angle, respectively, and the absorption value was represented in different colors. **c** and **e** plot the absorption lines at selected incident angles of 0, 30°, 50° and 70°. Simulated electric- (**f**, **h**, **j** and **l**) and magnetic-field (**g**, **i**, **k**, **m**) distributions of wide-band MMPA at different frequencies. The field patterns were obtained on the central cross section of unit cell. The *top to the bottom row* correspond to the working frequency of 0.99, 1.21, 1.32 and 1.48 THz, respectively. The normalized field magnitude is presented in different colors which have the same definition as those used in **b** and **d** [14]

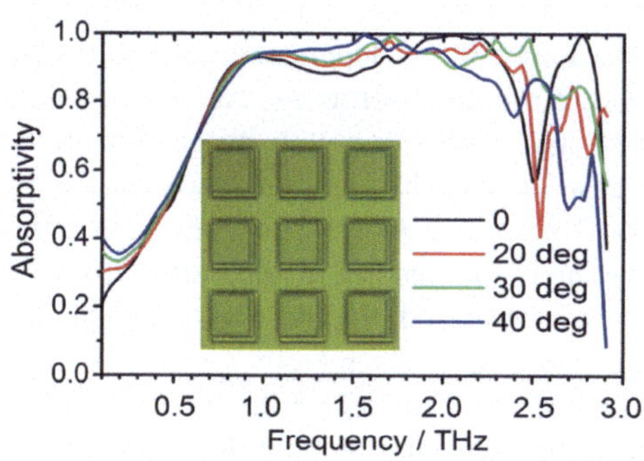

Fig. 5.23 Measured absorption of broadband THz MMPA at four incident angles: 0, 20°, 30°, and 40°. The *inset* gives a top view of the sample whose parameters were given in Fig. 3.34 and its caption [14]

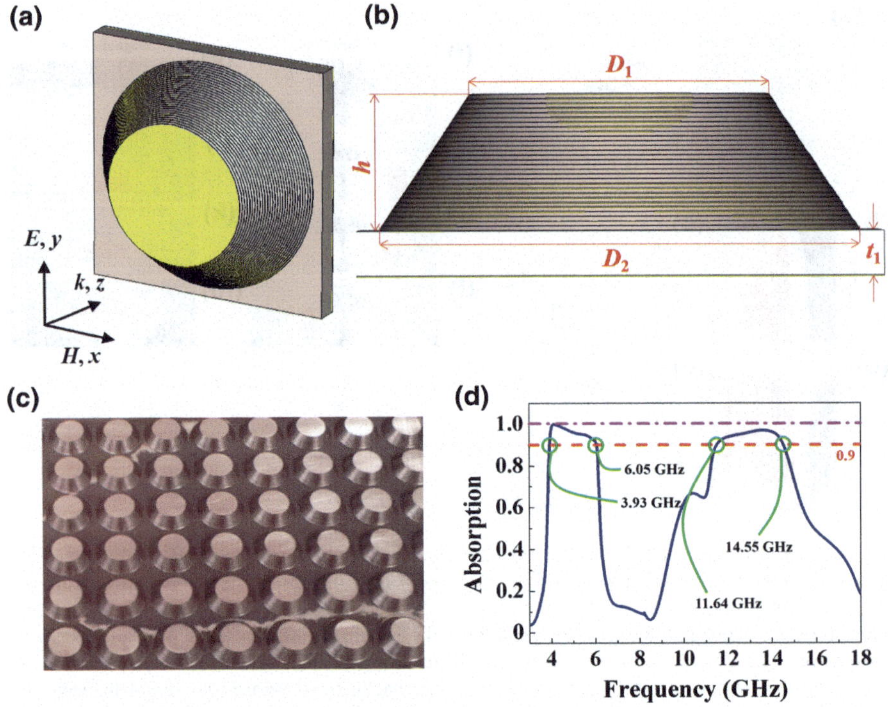

Fig. 5.24 **a** Perspective view and **b** side view of the designed meta-atom. D_1 and D_2 are the top and the bottom diameters of circular structures. t_1 and h are the thickness of FR-4 and the height of truncated-cone structure, respectively. **c** Photo of the fabricated sample. **d** Simulated absorption spectrum of the truncated cone-structure absorber [15]

reflection and absorption of the hybrid MM absorber with a conductivity of 0.02 S/m (the insulating state of VO_2 film). Two distinct absorption peaks appeared around 9.03 and 17.6 GHz with absorption of 94.9 % and 98 %, respectively. Figure 5.27a, b show the absorption spectra for the hybrid MM absorber according to temperature, and the temperature dependence of the absorption and the corresponding peak frequency for both absorptions. At room temperature, the absorption peaks occurred at 9.36 and 18.6 GHz with absorption of 87.0 % and 93.0 %, respectively. In the case of low-frequency response, the peak absorption was reduced slightly from 87.0 % at 9.36 GHz to 71.7 % at 9.98 GHz. On the other hand, a remarkable decline in the maximum absorption from 93.0 % at 18.6 GHz to 39.4 % at 19.1 GHz was observed for the high frequency. Figure 5.27c is the absorption spectra according to the conductivity of VO_2 film. By increasing the value of conductivity from 2–50 S/m at high frequency and from 0.02 to 2000 S/m at low frequency, the high-frequency absorption began to decrease from 93.8 to 19 % but the low-frequency one is reduced at the peak from 93.2 to 70.8 %. By varying the conductivity of VO_2 film, all features observed experimentally could be copied by simulation.

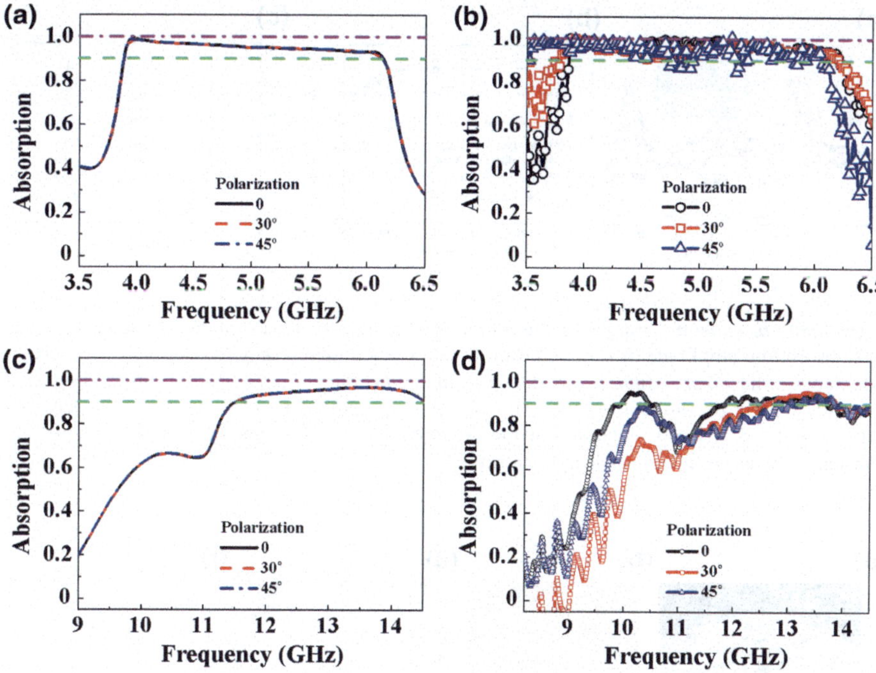

Fig. 5.25 **a**, **c** Simulated and **b**, **d** measured absorption spectra in 3.5–6.5 and 9.0–14.5 GHz [15]

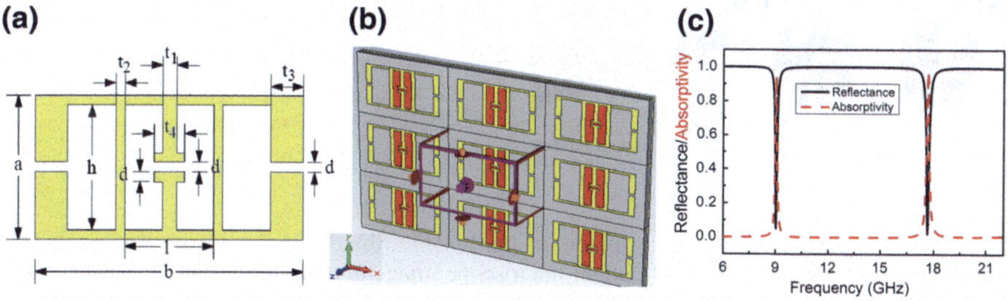

Fig. 5.26 Dual-band microwave MMPA. **a** Schematic of the designed eSRR unit cell and **b** perspective view of the absorber sheet with VO_2 patches. *Grey*, *white* (*yellow* in colour) and *black* (*red* in colour) represent the Al_2O_3 substrate, copper metal and VO_2 film, respectively. **c** Simulated reflectance (*black solid line*) and absorption (*red dashed line*) of the absorber [16]

The experimental demonstration of electronically-tunable MM absorbers in the THz regime was reported by Shrekenhamer et al. [17]. In Fig. 5.28a–d, a schematic illustrates the tuning mechanism by the liquid crystal. A potential was employed between electric-ring resonator (ERR) and ground plane, which orients the liquid crystal along the field lines. In Fig. 5.28, the frequency location of the absorption maximum (A_{max}) as a function of V_{bias} for modulation frequencies of 373 Hz,

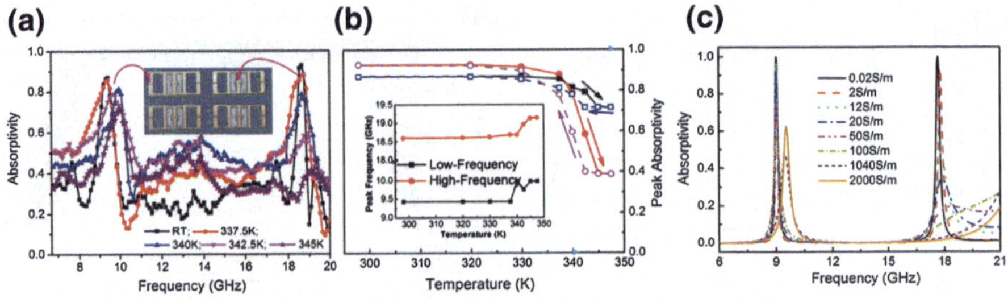

Fig. 5.27 **a** Measured absorption curves of the VO_2-based MMPA with respect to the device temperature. *Inset* is the image of the fabricated device. **b** Temperature dependence of the peak-absorption amplitude for both absorptions, where *solid symbols* stand for increase in temperature, and *empty symbols* for decrease in temperature. *Inset* is the corresponding peak frequency with respect to temperature. All *lines* were drawn as a guide to the eyes. **c** Dependence of the simulated absorption curves on the conductivity of VO_2 film [16]

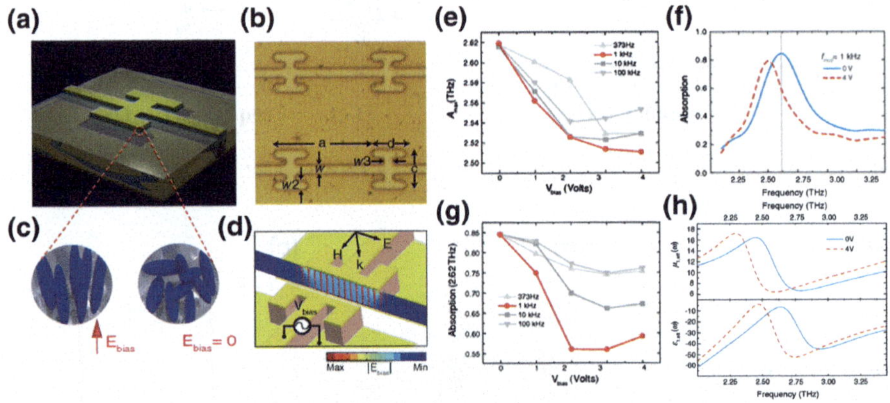

Fig. 5.28 Design and operational principle of the tunable MMPA. **a** Rendering of a single unit cell of the liquid crystal MMPA. **b** Optical-microscope image of a portion of the MM array where the unit cell had the dimensions of a = 50, c = 20, d = 16, w = $w2$ = 4.5, and $w3$ = 5 μm. **c** Depiction of the random alignment of liquid crystal in the unbiased case (*right*) and for an applied ac bias (*left*). **d** Simulation of the electric-field vector and the absolute value produced from an applied potential bias between ERR and ground plane. Experimentally-measured absorption of the MMPA. **e** Frequency location of the absorption maximum (A_{max}) as a function of applied bias voltage (V_{bias}) for modulation frequency (f_{mod}) of 373 Hz, 1 kHz, 10 kHz and 100 kHz. **f** Frequency-dependent absorption A(ω) for 0 (*blue solid curve*) and 4 V (*red dashed curve*) at f_{mod} = 1 kHz. *Solid line* is centered at $A_{max}(V_{bias} = 0)$ = 2.62 THz. **g** Absorption value at 2.62 THz as a function of V_{bias} for various modulation frequencies. **h** Effective optical constants of MMPA. Real part of the effective permeability ($\mu_{1,\ eff}$), and real part of the effective permittivity ($\varepsilon_{1,\ eff}$), determined from inversion of the simulated scattering parameters for 0 (*blue solid curves*) and 4 V (*red dashed curves*) [17]

1 kHz, 10 kHz, and 100 kHz was shown. The greatest frequency shift appeared for f_{mod} = 1 kHz, which was coincident with prior investigations with liquid crystal 4′-n-pentyl-4-cyanobiphenyl (5CB). The absorption peak appeared at 2.62 GHz with absorption of 85 % at V_{bias} = 0 and at 2.5 GHz with absorption of 80 %. A shift in the peak of the absorption by 4.6 % in frequency and narrower bandwidth were observed. The greatest change in $A(\omega_0)$ was obtained for 1-kHz bias modulation, as shown by the red-circle curve in Fig. 5.28e, which was coincident with the results presented in Fig. 5.28g. In Fig. 5.28h, both the permittivity and permeability moved with little change in their shape for all applied biases investigated. The demonstration of the dynamic control of EM waves at surfaces provided a new path of exotic devices.

Zhao et al. [18] numerically and experimentally reported the polarization-insensitive tunable MMPA with varactor diodes embedded between the unit cells. The structure was composed of two identical electric-field coupled-*LC* resonators with the same orientation on the top of dielectric substrate, which were connected by microwave varactor diode (see Fig. 5.29a, b). Figure 5.29c shows the simulated absorption of the designed structure according to reverse bias voltage. The results indicated that the tunable bandwidth turned out to be 1.19 GHz with the absorption over 99 %. The absorption frequency shifted from 4.45 to 5.64 GHz as the reverse bias voltage was tuned from 0 to −19 V. For the measured absorption, the

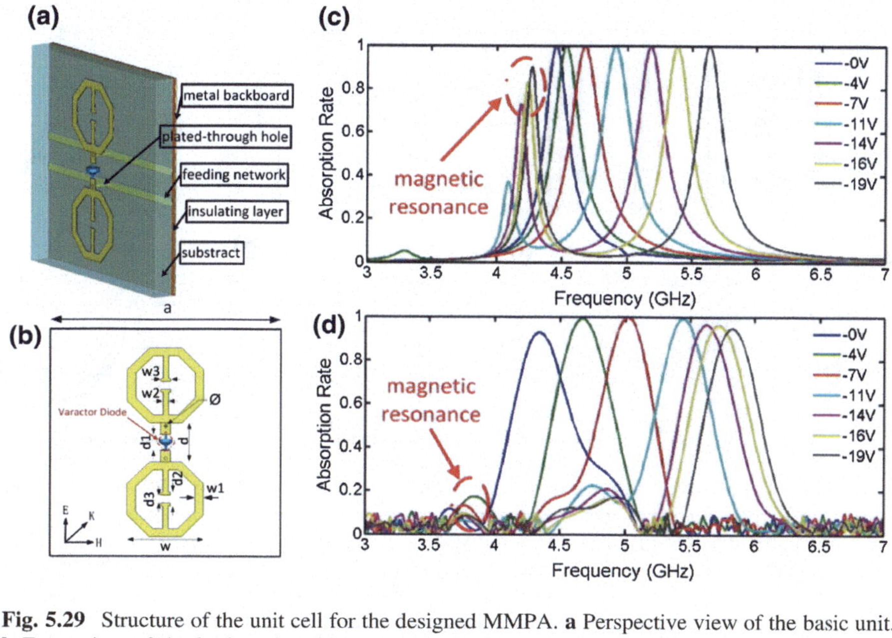

Fig. 5.29 Structure of the unit cell for the designed MMPA. **a** Perspective view of the basic unit. **b** Front view of the basic unit, with a varactor diode connecting the adjacent electric-*LC* units. **c** Simulation results of absorption at different reverse bias voltages, ranging from 0 to −19 V. **d** Experimental results of the absorption at different reverse bias voltages, ranging from 0 to −19 V. The *dashed circle* indicates the occurrence of the magnetic resonances [18]

absorption bandwidth was 1.5 GHz (from 4.35 to 5.85 GHz) with the absorption over 90 % as the bias voltage was tuned from 0 to −19 V (as shown in Fig. 5.29d). The measured bandwidth was slightly wider than the simulation.

Huang et al. [19] demonstrated the magnetically-biased frequency-tunable absorption by integrating a ferrite as the substrate or superstrate into the conventional planar MMPA. An array of metallic ELC resonators was backed up with a metallic mirror to realize the conventional passive MMPA. In order to achieve tunable characteristics, two strategies by integrating ferrite into the passive MMPA were intended (see Fig. 5.30a–c). One (MMPA-1) was that the ferrite layer was put between the dielectric layer and the ground plane. The other (MMPA-2) was that the ferrite layer was used as a superstrate, and another dielectric FR-4 layer was placed between the ferrite and the passive MMPA. Figure 5.30d–g show that the absorption peaks blue-shifted for both MMPAs by increasing gradually the magnetic field. The shift rates of absorption frequency of MMPA-1 and MMPA-2 were about 0.36 MHz/Oe and 0.18 MHz/Oe, respectively. When the measured results and the corresponding simulations were under the same magnitudes of magnetic fields, the resonance frequencies for both MMPAs turned out to be higher for the measurement compared with the corresponding calculation. A thicker thickness of FR-4 or ferrite layer, a stronger magnetic field was needed to ensure a nearly

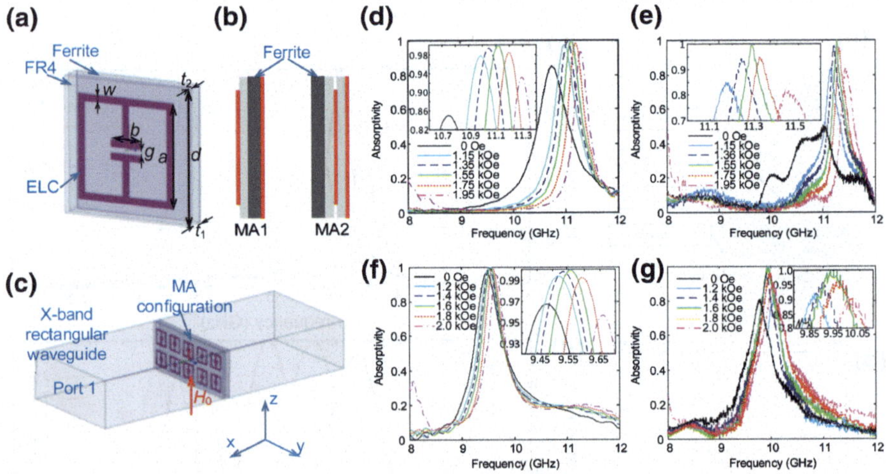

Fig. 5.30 **a** Unit cell of the ferrite-based tunable MMPA, **b** side views of two MMPA models, and **c** schematic representation of the measurement. **d, f** Simulated and **e, g** measured absorption properties under different magnetic fields for MMPA-1 and MMPA-2, respectively. *Insets* show the zoom-in plots. In detail, as the increase of magnetic field, the resonant frequency of MMPA-1 blue-shifted from 10.99 to 11.29 GHz in the simulation (from 11.17 to 11.45 GHz in measurement). The absorption first increased from 97.8 to 100 % and then descended to 93.3 % in the simulation (from 85 % to the unity and then down to 80 % in measurement). For MMPA-2, the resonant frequency blue-shifted from 9.52 to 9.66 GHz in the simulation (from 9.9 to 10.05 GHz in measurement), and the absorption first increased from 99.5 % to 100 % and then descended to 95.7 % in the simulation (from 92.5 to 100 % and then down to 89.9 % in measurement) [19]

uniform absorption, and the magnetic-field range for high absorption over 98.5 % was decreased. The shift rate of the resonant frequency was nearly linear when the magnetic field was relatively weak, and was strongly nonlinear by increasing the magnetic field, especially, for the MMPA-2. For a thinner FR-4 layer, the resonance frequency of the MMPA was more sensitive to the magnetic field, but unfortunately the high-absorption band was decreased as well (see Fig. 5.31a–d). Figure 5.31e presents that the magnetic field was changed from 0 to 1.9 kOe. The resonance frequency of the MMPA shifted from 8.6 to 9.65 GHz, and the absorption at each resonance remained higher than 98.5 %.

A plasmonic MM absorber for UV frequency with slight angle sensitivity was investigated by Hedayati et al. [20]. Figure 5.32a, b are the elemental map of the fabricated sample measured by high-resolution transmission-electron microscopy, and the schematic geometry of the designed plasmonic-MM super-absorber. The stack consisted of 4 layers: glass at the bottom, 200-nm silver film, 10–30 nm SiO_2, and finally 20 nm silver-SiO_2 composite. The gap between nanoparticles was very small owing to high filling factor of particles. Figure 5.32c, d show that the current plasmonic-MM super-absorber was nearly angle and polarization invariant through

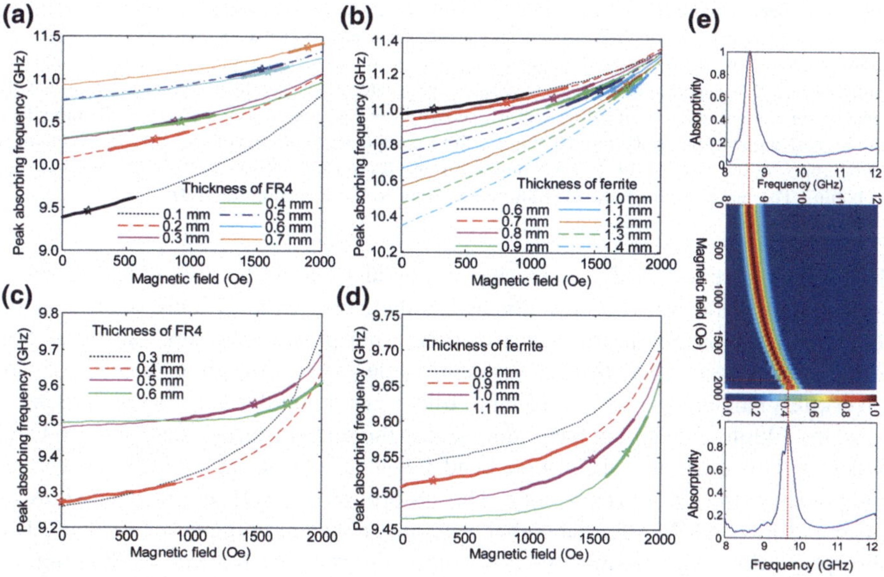

Fig. 5.31 Simulated resonant frequency versus magnetic field for MMPA-1 when the thicknesses of **a** FR4 and **b** ferrite layers were changed, and for MMPA-2 according to the thickness of **c** FR4 and **d** ferrite layers. The *star* and *thicker part* at each curve corresponds to the nearly uniform absorption, and the band of absorption was larger than 98.5 % for each condition. **e** Simulated absorption spectra of MMPA-2 under different magnetic fields with the optimized dimensional parameters. The *middle plot* presents the absorption spectra as function of magnetic field and frequency. *Left* and *right plots* show the absorption at magnetic field of 0 and 1900 Oe, respectively [19]

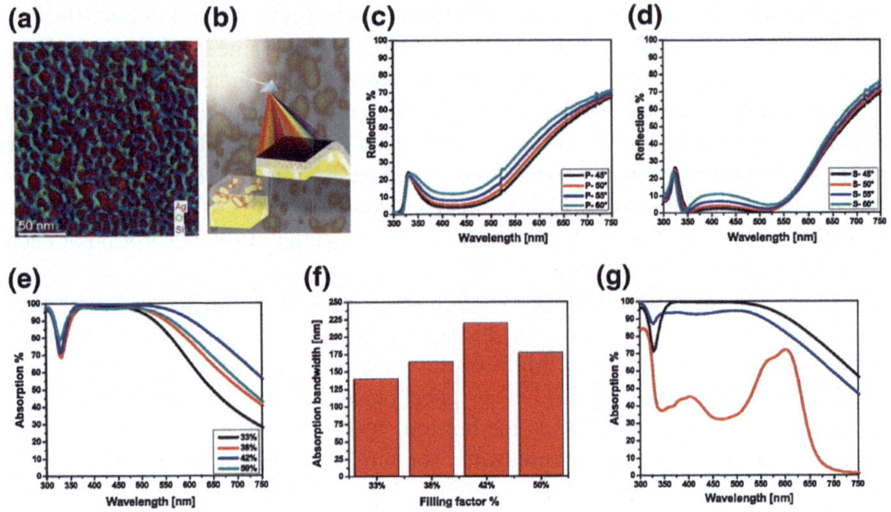

Fig. 5.32 **a** TEM elemental map of the optimized silver-SiO_2 nanocomposite for perfect absorption. **b** Schematic of the geometry of the MMs where the base layer was an optically-thick silver covered with 15-nm SiO_2 film. Atop, 15-nm Ag-SiO_2 composite with near-percolation filling factor was deposited. Reflection spectra of 20-nm silver-SiO_2 nanocomposite with 30 % filling factor deposited on 20-nm SiO_2-coated silver film (200-nm thick). **c** p and **d** s polarization at different angles of incidence. **e** Absorption spectra of silver-SiO_2 nanocomposite with different filling factors. **f** Absorption bandwidth with intensity greater than 95 % for different plasmonic-MM super-absorber with variety of filling factors, shown in (**e**). **g** Absorption spectra of 200-nm silver film coated with 40-nm polystyrene, that is, SPO composite (*red curve*) and 20-nm silver-SiO_2 nanocomposite with 42 % filling factor deposited on 10-nm (*blue curve*) and 15-nm (*black curve*) SiO_2 film. The organic film was UV-illuminated prior to the measurement [20]

that measured reflection by increasing the incident angle for two different polarizations. The reflection was low for both polarization even at high incident angle (60°), proving that the performance of such a plasmonic-MM super-absorber was not influenced significantly by incident angle. The absorption spectra of the silver-SiO_2 nanocomposite according to filling factor and the other parameters were constant. While the filling factor increased, the spectra red-shifted and its band widened. However, the absorption band began to shrink again over the certain composition (the optimized value). The absorption as well as the bandwidth of silver plasmonic-MM super-absorber was larger than its organic equivalent. The absorption dip at high frequency of this super-absorber was reduced by thinning the spacer layer to be 10 nm but in the expense of whole absorption (see Fig. 5.32e–g).

Yoo et al. [21] theoretically and experimentally reported a broadband MMPA by using water droplets. The unit cell was composed of water-droplet layer and metallic layer at the bottom, which were separated by the dielectric substrate layer. Figure 5.33a presents the pattern of water droplets which were periodically formed on flame retardant number 4 (FR-4: $\varepsilon = 4.3$ and $\tan\delta = 0.025$), the thickness of FR-4 was 2mm. During the diameter of water droplet was maintained at 8 mm, the

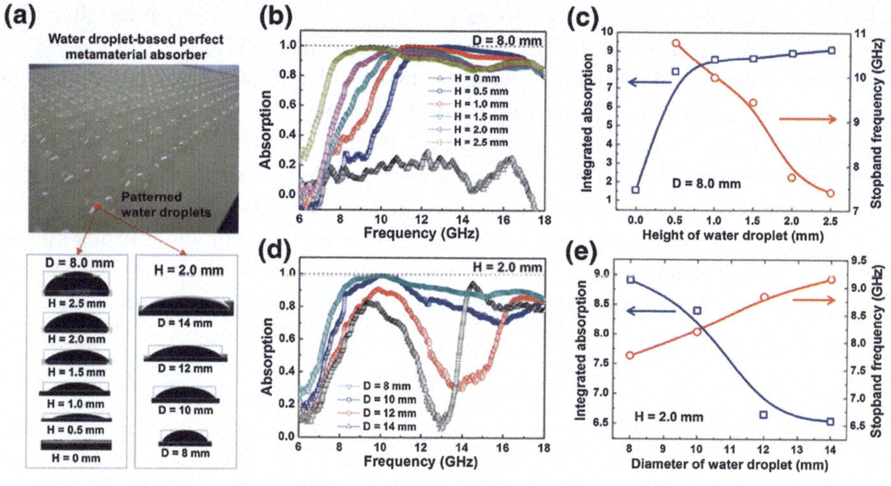

Fig. 5.33 **a** Images of FR-4 substrate with water droplets patterned periodically and contact-angle images of water droplet whose height and diameter are controlled. **b** Change of the absorption spectrum by changing the height of water droplet in 6 steps (0, 0.5, 1.0, 1.5, 2.0 and 2.5 mm) with fixed diameter of water droplet of 8.0 mm. **c** Total absorption and change of the stopband edge in 8–18 GHz by varying the water-droplet height. **d** Change of the absorption spectrum when the diameter of water droplet is varied in 4 steps (8, 10, 12 and 14 mm) with fixed height of water droplet of 2.0 mm. **e** Total absorption and change of the stopband edge in 8–18 GHz by varying the water-droplet diameter [21]

height of water droplet was controlled in 6 steps from 0 to 0.5, 1.0, 1.5, 2.0 and 2.5 mm. Figure 5.33b shows change of the absorption spectrum from no water to high water drops. Substrate without water droplet generates weak absorption, caused by dielectric loss. On the contrary, if the height is more than 0.5 mm, there is a characteristic of perfect absorber whose absorption is more than 93 %. Note that perfect absorption commonly means absorption over 93 %. It's interesting that the absorption spectrum is shifted to low-frequency domain according to the height of water. As the height of water was heighten to be 0.5, 1.0, 1.5, 2.0 and 2.5 mm, the perfect-absorption (over 93 %) frequency was transferred gradually to low-frequency range to be 11.50, 10.58, 10.29, 9.62 and 7.95 GHz. Especially, when the height of water droplet was 2.5 mm, more than 91 % of average absorption was shown in very wide range of $\Delta = 10$ GHz (shown in Fig. 5.33c). Furthermore, they also controlled the diameter of water droplet to be 8, 10, 12 and 14 mm. Figure 5.33d reveals change of the absorption spectrum according to the diameter of water droplet (14, 12, 10 and 8 mm, while the height of water droplet was fixed at 2.0 mm). The absorption was increased and the absorption range was also increased when the diameter of water droplet was reduced from 14 to 8 mm. Figure 5.33e presents the correlation between total absorption and stopband-edge frequency in 8–18 GHz by changing the water-droplet diameter. As the diameter of water droplet was reduced, total absorption was enhanced in 8–18 GHz. The total absorption quantity was increased to be 6.55, 6.7, 8.4 and 8.91 when the diameter of water was reduced to be 14, 12, 10 and 8 mm, respectively.

As aforementioned above, generally MMPAs have the high absorption at specific frequency and the narrow absorption band. To improve them, many researcher have investigated multi-band, broadband and tunable MMPAs. Multi-band MMPAs have been investigated by using the corresponding kinds of meta-patterns. Broadband MMPAs utilizing resist sheet, natural tap water and so on, are being actively investigated. Moreover, MMPAs based on tunable absorption and resonance frequency have also been in progress. All these researches are for real applications relevant to broadband and tunability in the near future.

References

1. N.I. Landy, S. Sajuyigbe, J.J. Mock, D.R. Smith, W.J. Padilla, Perfect metamaterial absorber. Phys. Rev. Lett. **100**, 207402 (2008)
2. Q.-Y. Wen, H.-W. Zhang, Y.-S. Xie, Q.-H. Yang, Y.-L. Liu, Dual band terahertz metamaterial absorber: design, fabrication, and characterization. Appl. Phys. Lett. **95**, 241111 (2009)
3. B. Zhang, Y. Zhao, Q. Hao, B. Kiraly, I.-C. Khoo, S. Chen, T.J. Huang, Polarization-independent dual-band infrared perfect absorber based on a metal-dielectric-metal elliptical nanodisk array. Opt. Express **19**, 15221 (2011)
4. X. Liu, T. Tyler, T. Starr, A.F. Starr, N.M. Jokerst, W.J. Padilla, Taming the blackbody with Infrared metamaterials as selective thermal emitters. Phys. Rev. Lett. **107**, 045901 (2011)
5. Z.H. Jiang, S. Yun, F. Toor, D.H. Werner, T.S. Mayer, Conformal dual-band near-perfectly absorbing mid-infrared metamaterial coating. ACS Nano **5**, 4641 (2011)
6. X. Shen, Y. Yang, Y. Zang, J. Gu, J. Han, W. Zhang, T.J. Cui, Triple-band terahertz metamaterial absorber: design, experiment, and physical interpretation. Appl. Phys. Lett. **101**, 154102 (2012)
7. H.-X. Xu, G.-M. Wang, M.-Q. Qi, J.-G. Liang, J.-Q. Gong, Z.-M. Xu, Triple-band polarization-insensitive wide-angle ultra-miniature metamaterial transmission line absorber. Phys. Rev. B **86**, 205104 (2012)
8. J.W. Park, P.V. Tuong, J.Y. Rhee, K.W. Kim, W.H. Jang, E.H. Choi, L.Y. Chen, Y.P. Lee, Multi-band metamaterial absorber based on the arrangement of donut-type resonators. Opt. Express **21**, 9691 (2013)
9. Y.J. Yoo, Y.J. Kim, P.V. Tuong, J.Y. Rhee, K.W. Kim, W.H. Jang, Y.H. Kim, H. Cheong, Y. P. Lee, Polarization-independent dual-band perfect absorber utilizing multiple magnetic resonances. Opt. Express **21**, 32484 (2013)
10. H. Wakatsuchi, S. Greedy, C. Christopoulos, J. Paul, Customized broadband metamaterial absorbers for arbitrary polarization. Opt. Express **18**, 22187 (2010)
11. L.K. Sun, H.F. Cheng, Y.J. Zhou, J. Wang, Broadband metamaterial absorber based on coupling resistive frequency selective surface. Opt. Express **20**, 4675 (2012)
12. Y. Cui, J. Xu, K.H. Fung, Y. Jin, A. Kumar, S. He, N.X. Fang, A thin film broadband absorber based on multi-sized nanoantennas. Appl. Phys. Lett. **99**, 253101 (2011)
13. D. Ye, Z. Wang, K. Xu, H. Li, J. Huangfu, Z. Wang, L. Ran, Ultra wideband dispersion control of a metamaterial surface for perfectly-matched-layer-like absorption. Phys. Rev. Lett. **111**, 187402 (2013)
14. J. Zhu, Z. Ma, W. Sun, F. Ding, Q. He, L. Zhou, Y. Ma, Ultra-broadband terahertz metamaterial absorber. Appl. Phys. Lett. **105**, 021102 (2014)
15. Y.J. Kim, Y.J. Yoo, K.W. Kim, J.Y. Rhee, Y.H. Kim, Y.P. Lee, Dual broadband metamaterial absorber. Opt. Express **23**, 3861 (2015)

16. Q.-Y. Wen, H.-W. Zhang, Q.-H. Yang, Z. Chen, Y. Long, Y.-L. Jing, Y. Lin, P.-X. Zhang, A tunable hybrid metamaterial absorber based on vanadium oxide films. J. Phys. D Appl. Phys. **45**, 235106 (2012). doi:10.1088/0022-3727/45/23/235106
17. D. Shrekenhamer, W.-C. Chen, W.J. Padilla, Liquid crystal tunable metamaterial absorber. Phys. Rev. Lett. **110**, 177403 (2013)
18. J. Zhao, Q. Cheng, J. Chen, M.Q. Qi, W.X. Jiang, T.J. Cuil, A tunable metamaterial absorber using varactor diodes. New J. Phys. **15**, 043049 (2013)
19. Y. Huang, G. Wen, W. Zhu, J. Li, L.-M. Si, M. Premaratne, Experimental demonstration of a magnetically tunable ferrite based metamaterial absorber. Opt. Express **22**, 16408 (2014)
20. M.K. Hedayati, A.U. Zillohu, T. Strunskus, F. Faupel, M. Elbahri, Plasmonic tunable metamaterial absorber as ultraviolet protection film. Appl. Phys. Lett. **104**, 041103 (2014)
21. Y.J. Yoo, S. Ju, S.Y. Park, Y.J. Kim, J. Bong, T. Lim, K.W. Kim, J.Y. Rhee, Y.P. Lee, Metamaterial absorber for electromagnetic waves in periodic water droplets. Sci. Rep. **5**, 14018 (2015)

Chapter 6
Polarization-Independent and Wide-Incident-Angle Metamaterial Perfect Absorber

Abstract The polarization of electromagnetic (EM) wave takes an important place in research of metamaterial perfect absorber (MMPA). In this chapter, we review the influence of the polarization of EM wave on MMPAs and, then, show polarization-independent MM absorbers by taking the advantage of structural symmetry. Even though the absorption is strongly reduced by increasing the incident angle of EM wave, to develop MM absorbers for the practical applications, we should use some special designs to reveal wide-incident-angle MMPAs. An analytical model is introduced by mean of the equivalent circuit in order to optimize the structure capable to work for larger incident angle. Finally, by introducing perfectly-matched-layer-like structure, MMPAs with wider incident angle and, at the same time, higher absorption can be achieved.

6.1 Introduction

The realization of a "material" that absorbs all radiation incident, regardless of frequency and/or angle of incidence, has been described a century ago by Planck's law [1]. However, none of the natural materials behave like an ideal blackbody. Instead, each material has its own radiation coefficient. For example, in some kinds of glass, the electromagnetic (EM) wave is transparent in the visible range, and largely opaque to ultraviolet wave. To meet the need of seeking for a material absorbing EM wave regardless of frequency range, metamaterials (MM) were rising as potential candidate, the so-called metamaterial perfect absorber (MMPA) was first exploited by Landy et al. [2], taking advance of this field, a great number of optimized MMPA have been proposed for different frequency ranges [3–6], and application areas [7–9]. In this chapter, within the framework of the polarization of EM wave, two problems are going to be improved. At first, by making use of structural design, polarization-independent and some wide-incident-angle MMPA were proposed. An analytical model is also provided to characterize the properties of these MMPA. Next, a bird eyes of perfectly-matched layer (PML) and uniaxial

Y.P. Lee et al., *Metamaterials for Perfect Absorption*,
Springer Series in Materials Science 236, DOI 10.1007/978-981-10-0105-5_6

PML (UPML) are introduced to open a way to access really incident-angle-independent MMPA. Finally, by modifying the condition of UPML, a PML-like structure is obtained to be polarization-independent and wide for incident angle.

6.2 Dependence of Electromagnetic Polarization on Metamaterial Perfect Absorber

To operate the MMPA in the practical applications, the polarization sensitivity is being one of the main concerns. Polarization-sensitive type of MMPAs cannot operate at all the polarization sates. They can work only at specific polarization of incident wave, and limit the applications where polarization independence is important. Therefore, the research on polarization-independent MPMAs is necessary in both theory and experiment. Generally, the polarization insensitivity for EM wave at the normal incidence can be easily attained by employing a symmetrical design of metallic arrays on the first layer of MMPAs. In this Section, we will consider some MMPA structures which reveal the polarization independence for the normal incident wave.

One of the first symmetric MMPAs is theoretically and experimentally demonstrated by Landy et al. in 2009 [10]. The proposed unit cell of polarization-independent MMPA for EM wave at normal incidence is shown in Fig. 6.1. By using the effective-medium model, they can independently control the real and imaginary components of effective permittivity and permeability of MMPA to achieve the high absorption. Generally, the absorption of material can be calculated as $A = 1 - R(\omega) - T(\omega)$, where $R(\omega)$ and $T(\omega)$ are transmission and reflection, respectively. For the complex transmissivity [$t(\omega)$] and reflectivity [$r(\omega)$], the absorption is given by: $A = 1 - |r(\omega)|^2 - |t(\omega)|^2$. The dependence of transmissivity on the complex index of refraction ($n = n_1 + in_2$) and impedance ($Z = Z_1 + iZ_2$) for a slab of length d is written by:

$$t(\omega)^{-1} = \left\{\sin(nkd) - \frac{i}{2}\left(Z + \frac{1}{Z}\right)\cos(nkd)\right\}e^{ikd}, \tag{6.1}$$

where the wave number $k = \omega/c$ and c is the speed of light in vacuum. When Z reaches the free-space value ($Z = 1$), the reflectivity is zero and the transmissivity can be determined by equation

$$t(\omega)^{-1} = \{\sin(nkd) - i\cos(nkd)\}e^{ikd} = e^{-i(n_1-1)kd}e^{n_2kd}. \tag{6.2}$$

Consequently, the transmission $T = |t|^2 = e^{-2n_2kd}$ and $T = 0$ when n_2 reaches infinity. Therefore, the maximum of absorption $A = 1$ when $T = R = 0$. In other words, in order to obtain the high absorption, it is necessary for $Z = 1$ at a point where n_2 is large. This statement indicates that the complex frequency-dependent

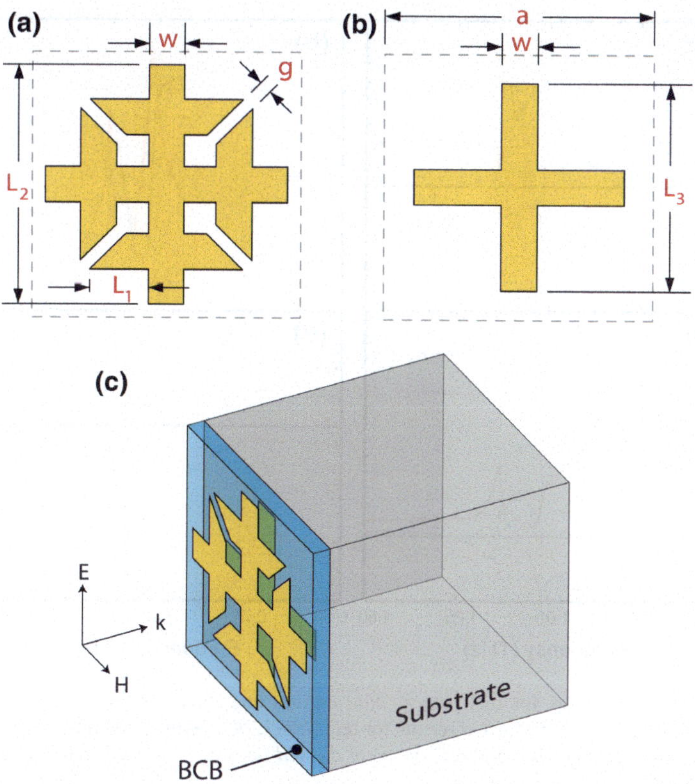

Fig. 6.1 Perspective view of **a** coupled ring resonator (ERR), **b** cross, and **c** combined ERR and cross. The geometrical parameters are a = 84, L_1 = 52.5, L_2 = 74, L_3 = 64, w = 11, and g = 4 μm [10]

permittivity [$\varepsilon(\omega)$] and permeability [$\mu(\omega)$] of a MMPA need to be finely controlled. The refractive index and the impedance can be represented by $n = \sqrt{\varepsilon(\omega)\mu(\omega)}$ and $Z = \sqrt{\mu(\omega)/\varepsilon(\omega)}$, respectively.

Ordinarily, the resonant phenomenon in MMPA is described in the form of a complex oscillator in frequency:

$$\varepsilon(\omega), \mu(\omega) = \varepsilon_\infty, \mu_\infty + \frac{F_{\varepsilon,\mu}\omega^2}{\omega_{0\varepsilon,\mu}^2 - \omega^2 - i\gamma\omega}, \tag{6.3}$$

where F is the oscillator strength, γ is the damping, ω_o is the center frequency of oscillator, and, ε, μ are high-frequency contributions. As shown in Fig. 6.1a, the coupled ring resonator (ERR) is designed to control the electric response. The polarization-insensitive property of proposed structure is designed by the fourfold rotational symmetry of ERR about the propagation axis. Simultaneously, the

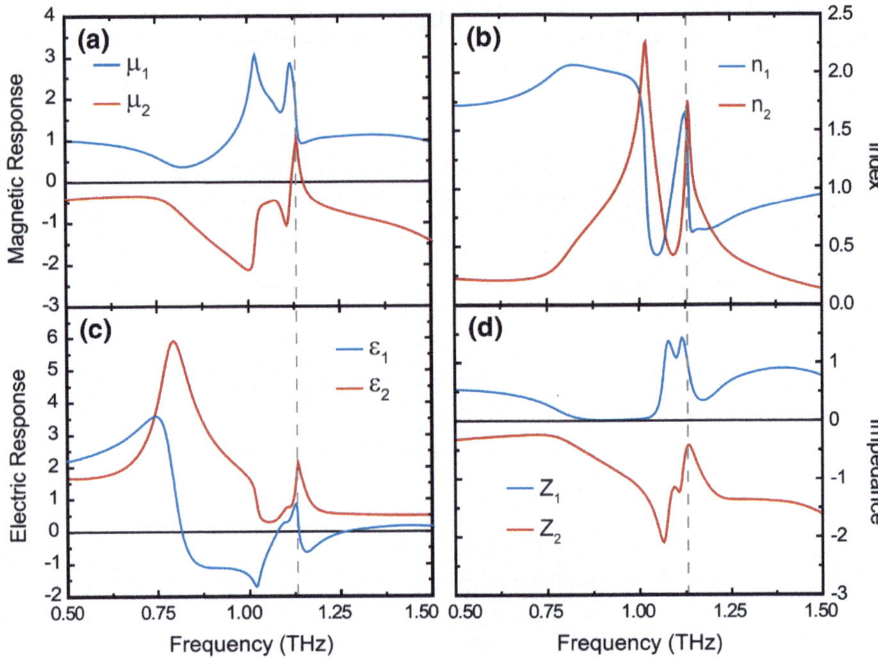

Fig. 6.2 Extracted results for real (*black, blue online*) and imaginary (*dark gray, red online*) components of **a** μ and **c** ε. Extracted results for real (*black, blue online*) and imaginary (*dark gray, red online*) components of **b** n and **d** Z. *Vertical dashed lines* display the frequency of maximum absorption [10]

magnetic response is controlled by coupling between the ERR with a cross structure, which is separated by the benzocyclobutane (BCB) spacer (Fig. 6.1b and c).

Figure 6.2 shows the extracted optical constants for the proposed MMPA. The real and the imaginary components of μ and ε are plotted as in Fig. 6.2a and c, respectively. It is noteworthy that the lowest-frequency feature in ε_2 at $\omega = 800$ GHz is the conventional-MM electrically-coupled inductive-capacitive resonance of the ERR (Fig. 6.2c). A second electric resonance appears around 1.125 THz due to the cut-wire response of the cross. The resonance in μ is weak relative to the neighboring resonances in ε and thus difficult to observe. However, μ_2 has a distinct positive peak at 1.13 THz. The refractive index n and the impedance Z can be calculated as presented in Fig. 6.2b and d. As in Fig. 6.2b, the imaginary index n_2 is maximized to be 1.75, which minimizes $T = 0$. The real impedance $Z_1 = 1$, and the complex impedance Z_2 are minimized, leading to $R = 0$ at 1.13 THz (Fig. 6.2d). These results create an absorption peak of 95 % at 1.13 THz as shown in Fig. 6.3 (gray-green solid curve online).

In order to confirm the simulated results, the corresponding experiment was performed as shown in Fig. 6.4. The best fit of simulation to the experimental data (for measured T and R) was used to obtain the experimental absorption A. In

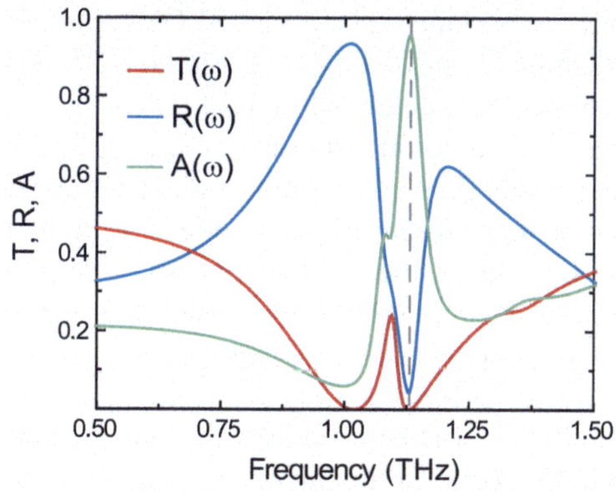

Fig. 6.3 Simulated reflection (*black, blue online*), transmission (*dark gray, red online*) and absorption spectrum (*light gray, green online*) for the proposed MMPA. *Vertical dashed line* displays the frequency of maximum absorption [10]

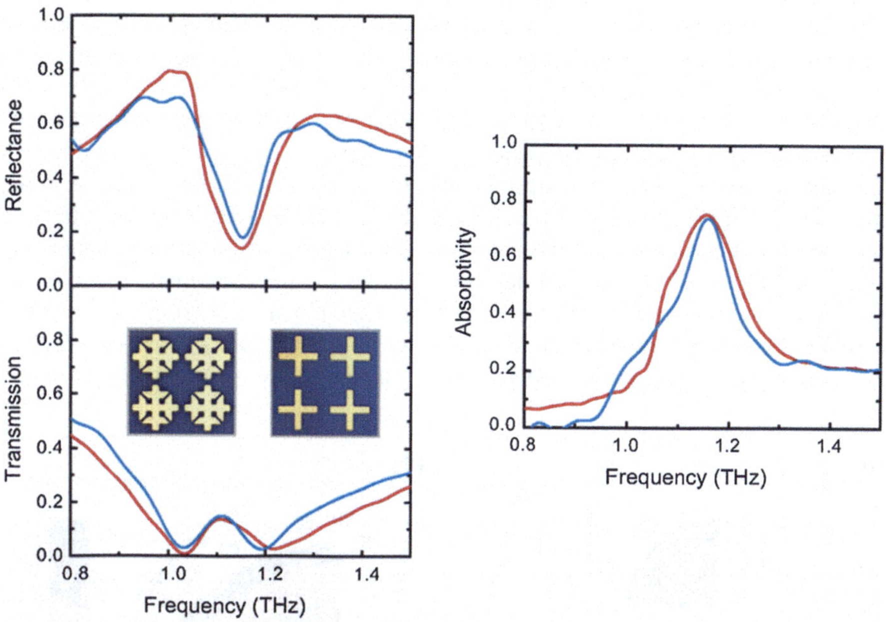

Fig. 6.4 (*Left*) Reflection and transmission spectra by experiment (*black, blue online*) and simulation (*dark gray, red online*). *Insets* are two unit cells (one for each layer) of real sample. (*Right*) Experimental (*black, blue online*) and simulated (*dark gray, red online*) absorption spectra of the proposed MMPA [10]

comparison to the theoretical results in Fig. 6.3, the fabricated sample exhibits a shift in the resonant frequency due to small fabrication tolerance for the structure. It is also noteworthy that the first transmission minimum is shifted by about 10 GHz, while the second minimum is changed by about 54 GHz, both to higher frequencies. The minimum in reflection is shifted from 1.128 to 1.15 THz. Consequently, a measured absorption of 65 % is achieved at 1.145 THz.

Among many types of symmetric structures, the ring resonators were also particularly studied for the single/dual polarization-independent MMPA. By exploiting two concentric circular-shaped structures embedded in a single unit cell (as shown in Chap. 3), Tuong et al. introduced the sandwiched structure (metal-dielectric-metal) for dual-band perfect absorption [11]. For this model, the bottom layer of MMPA is replaced by the continuous metallic layer, which completely blocks all incident EM waves. Therefore, the transmission is zero ($T = 0$) and the absorption is expressed as $A = 1 - R(\omega)$. The dual-band MMPA exhibits the polarization-insensitive properties which are plotted in Fig. 6.5a for the microwave range (8.55 and 11.78 GHz) and in Fig. 6.5b for the mid-infrared (IR) region (11.8 and 17.9 THz). It could be found that the absorption spectrum was unchanged according to polarization angle ϕ from 0 to 90° for the normal incident wave.

With a set of split rings and complete rings to be a combined structure, they could also obtain the polarization-independent property of MMPA. Ghosh et al. introduced the design including an array of interconnected rings [12]. Two sets of concentric rings were embedded one inside another, where each set, comprising of one split ring and one complete ring, produced dual-band perfect absorption. The proposed MMPA with a periodic arrangement of the unit cell and its equivalent-circuit model are depicted in Fig. 6.6. For the oscillation circuit in Fig. 6.6b, L_1 and C_1 were related to the effective inductance and capacitance of the inner set of the unit cell, whereas L_2 and C_2 were regarded as the effective inductance and capacitance of the outer set. These equivalent-circuit parameters L_1, L_2, C_1, and C_2 have been divided into two separate parallel combinations according to current-flow direction in the top frequency-selective surface (FSS). C_P was the coupling capacitance between inner and outer unit cell and C_D was the effective dielectric capacitance between top and bottom metallic layers. For simplicity in this

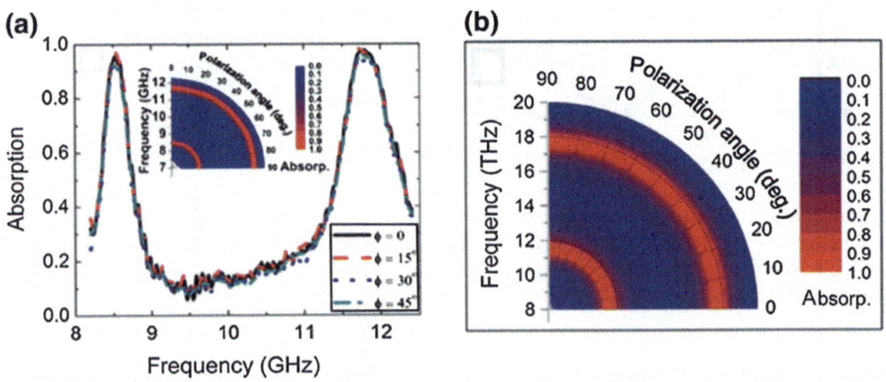

Fig. 6.5 **a** Absorption spectra according to polarization angle (simulated result in the *inset*). **b** Results of polarization-insensitive dual-band absorption in the mid-IR region [11]

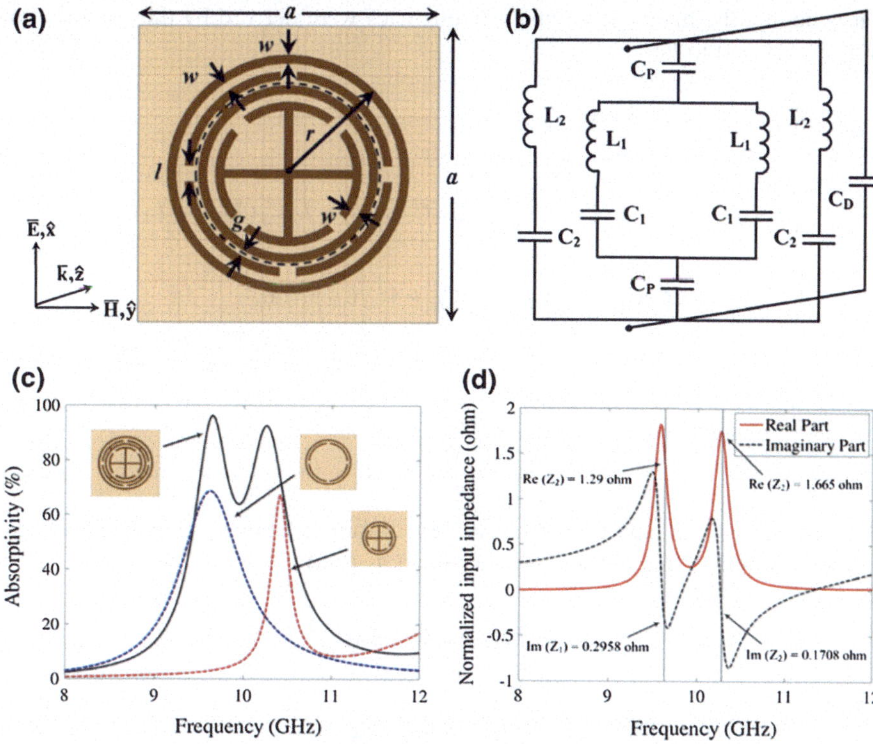

Fig. 6.6 **a** Front view of the unit-cell structure with geometrical dimensions: a = 7.2, r = 2.89, w = 0.2, g = 0.14, and l = 0.4 mm. **b** Equivalent *RLC* circuit of the proposed structure. **c** Simulated absorption spectra in order to explain the mechanism of dual-band perfect absorption. **d** Calculated normalized input impedance of the proposed structure [12]

case, the resistances (defining the Ohmic and the dielectric loss) and the mutual inductance (between inner and outer set of unit cells) were ignored.

In order to calculate the absorption frequencies of the proposed absorber, they found the equivalent impedance of the circuit $Z(\omega)$ as (6.4):

$$Z(\omega) = \left[\left(j\omega\frac{L_2}{2} + \frac{1}{j\omega(2C_2)}\right) \middle\| \left(j\omega\frac{L_1}{2} + \frac{1}{j\omega(2C_1)} + \frac{1}{j\omega(C_P/2)}\right)\right] \middle\| \left(\frac{1}{j\omega C_D}\right). \tag{6.4}$$

After some algebraic manipulations, $Z(\omega)$ could be rewritten as

$$Z(\omega) = \left[\frac{(1-\omega^2 L_2 C_2)\times\left(\frac{C_P}{2} + 2C_1 - \omega^2 L_1 C_1 \frac{C_P}{2}\right)}{j\omega C_1 \times C_P \times (1-\omega^2 L_2 C_2) + \omega \times 2C_2 \times \left(\frac{C_P}{2} + 2C_1 - \omega^2 L_1 C_1 \frac{C_P}{2}\right)}\right] \middle\| \left(\frac{1}{j\omega C_D}\right). \tag{6.5}$$

Consequently, the magnetic resonant frequencies were derived by the condition of $Im(1/Z(\omega)) = 0$ to be

$$\omega^4 \times \left(L_1 L_2 C_1 C_2 C_D \frac{C_P}{2}\right) - \omega^2 \times \left(L_1 C_1 C_D \frac{C_P}{2} + L_2 C_2 C_D \frac{C_P}{2} + 2L_2 C_1 C_2 C_D + L_2 C_1 C_2 C_D + L_1 C_1 C_2 C_P\right) + \left(C_D \frac{C_P}{2} + 2C_1 C_D + C_1 C_P + C_2 C_P + 4C_1 C_2\right) = 0, \tag{6.6}$$

or

$$\omega^4 \times A - \omega^2 \times B + C = 0, \tag{6.7}$$

where constants A, B, and C could be achieved from the above equation. Therefore, the two absorption frequencies were obtained as (6.8).

$$f_1 \approx \frac{1}{2\pi}\left(\frac{B + \sqrt{B^2 - 4AC}}{2A}\right)^{\frac{1}{2}}, \quad f_2 \approx \frac{1}{2\pi}\left(\frac{B - \sqrt{B^2 - 4AC}}{2A}\right)^{1/2}, \tag{6.8}$$

where the higher (lower) absorption frequency was due to the inner (outer) set of unit cells (as shown in Fig. 6.6c). The combined structure had two absorption peaks at 9.66 and 10.26 GHz (with absorption of 96 and 92.5 %, respectively). The calculated surface impedance $Z(\omega)$ matched closely to the free space impedance Z_0 as in Fig. 6.6d, which confirmed more clearly about the mechanism of dual-band MMPA. It was observed that the real parts of $Z(\omega)$ were close to unity, while the imaginary parts of $Z(\omega)$ were nearly zero for both absorption peaks at 9.66 and 10.26 GHz. Due to the symmetrical design, the polarization insensitivity was established for all polarization angles under the normal incidence in Fig. 6.7a for simulation and Fig. 6.7b for experiment. It was also noted that this MMPA model could be applied for the target of enhanced bandwidth of operation spectrum by controlling two absorption peaks to be partly overlapped.

Besides, another work by S. Bhattacharyya and co-worker studied a triple-band polarization-independent MM absorber using electric field-driven *LC* (ELC) resonators [13]. From the asymmetric ELC resonator, the diagonal symmetry of triple-band MMPA was constructed by rotating the single unit cell by 90°. The proposed MMPA was a periodic arrangement of the unit cell depicted in Fig. 6.8a, where sub-unit cell “2” was rotated by 90° with respect to sub-unit cell “1” of the same dimensions. Figure 6.8b show the simulated absorption peaks at 4.74, 5.50,

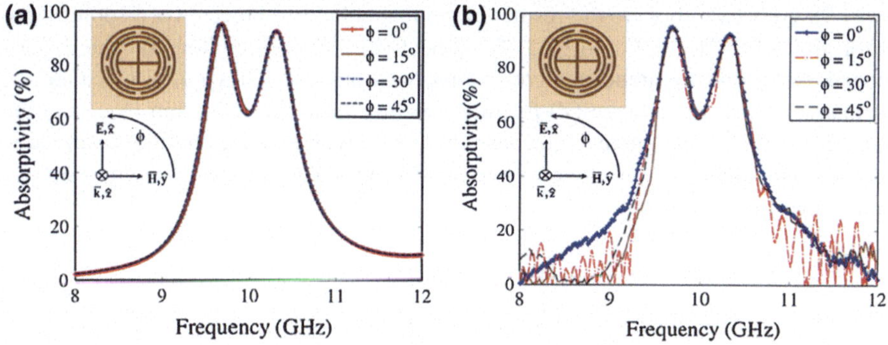

Fig. 6.7 Polarization-independent absorption spectra under normal incidence for **a** simulation and **b** measurement [12]

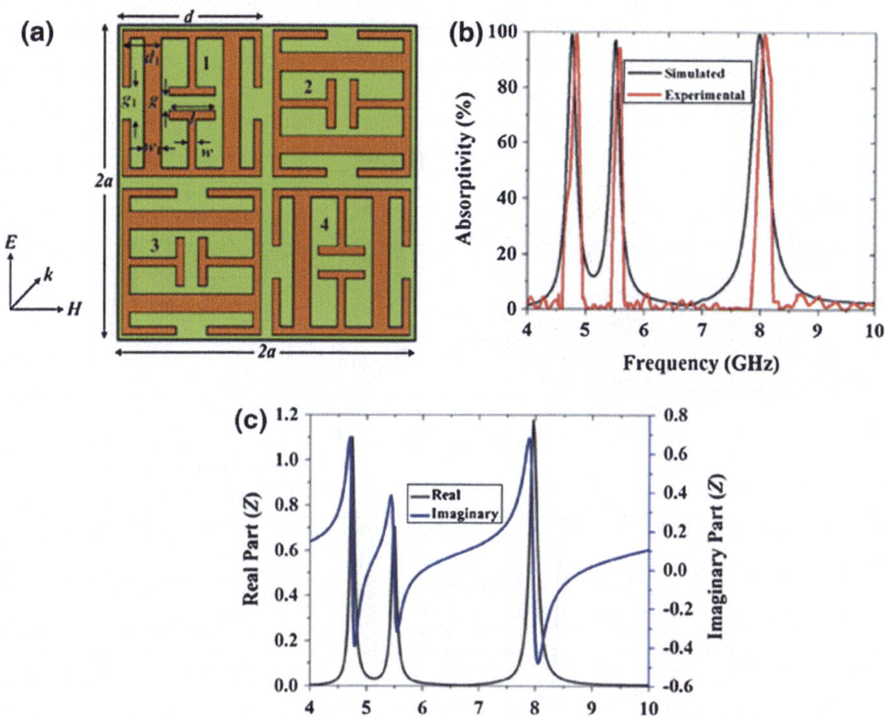

Fig. 6.8 **a** Unit cell, and **b** comparison between simulated and measured absorption spectra of the proposed triple-band MMPA. The optimized parameters of structure: 2a = 18, d = 8.05, $d_1 = 2.0125$, w = 0.5, $w_1 = 1.2$, g = 0.8, $g_1 = 1.7$, and l = 3.45 mm. **c** Response of the extracted normalized input impedance (Z) from (6.9) [13]

and 7.98 GHz (which were the results of the magnetic resonances) corresponding to absorption of 98.9, 96.9, and 99.1 %, respectively. The simulated and the experimental absorption frequencies were relatively coincident. They could calculate the normalized input impedance (Z) for this structure from the reflection (S_{11}) and the transmission (S_{21}) parameters. Because the back layer was the continuous metallic plane, the incident EM wave was prevented from transmitting (S_{21}) = 0, the surface impedance became to be:

$$Z = \sqrt{\frac{(1+S_{11})^2 - S_{21}^2}{(1-S_{11})^2 - S_{21}^2}} = \frac{1+S_{11}}{1-S_{11}}. \tag{6.9}$$

As shown in Fig. 6.8c, the real and the imaginary parts of Z were close to unity and zero, respectively, for all the resonant frequencies (4.74, 5.50 and 7.98 GHz), implying no reflection from the surface. The influence of the polarization of EM wave on absorption is shown in Fig. 6.9. Due to the structural symmetry in the ELC plane (Fig. 6.9a), the structure was investigated by changing the polarization angle from 0 to 45° for normally-incident wave. As observed in Fig. 6.9b and c, the

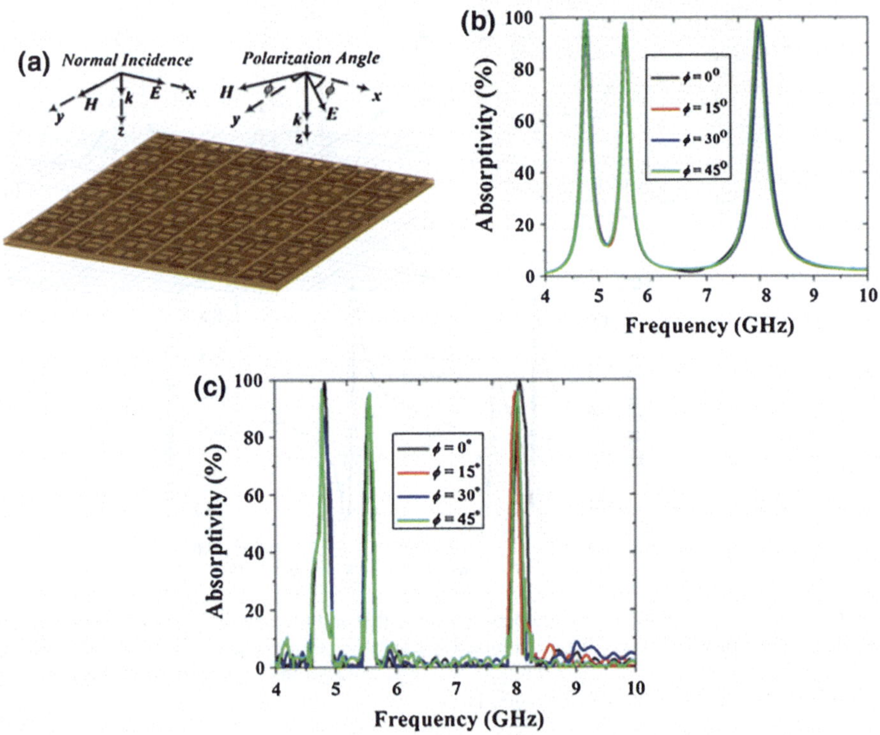

Fig. 6.9 **a** Arrangement of the structure for variation of the polarization angle (ϕ). **b** Simulated and **c** measured response of the structure at various polarization angles [13]

simulated and the experimental absorption spectra were unchanged for all of polarization angles. This phenomenon confirmed the polarization-independent nature of proposed MMPA.

In order to be expanded to practical application areas, we also need to develop the broadband polarization-independent MMPA. These results are promising for solar cells, stealth, and imaging applications. To follow this way, Wang and co-workers suggested a simple method for significantly broadened absorption bandwidth of MMPA based on sandwiched structure: two low-conductivity square alloy rings on top and ground layer, separated by a dielectric layer [14]. This proposed alloy absorber was polarization-insensitive owing to its symmetric structure. The unit cell of the compact alloy absorber structure is presented in Fig. 6.10.

As shown in Fig. 6.10c, two magnetic resonances induced two absorption peaks at 2.00 (f_1) and 4.33 (f_2) THz with absorption of 99.25 and 99.73 %, respectively. Because these two absorption peaks were partly overlapped, they could obtain the broad-band absorption with bandwidth (absorption higher than 50 %) of 3.96 THz (from 1.37 to 5.33 THz). The relative full width at half maximum (FWHM) of the

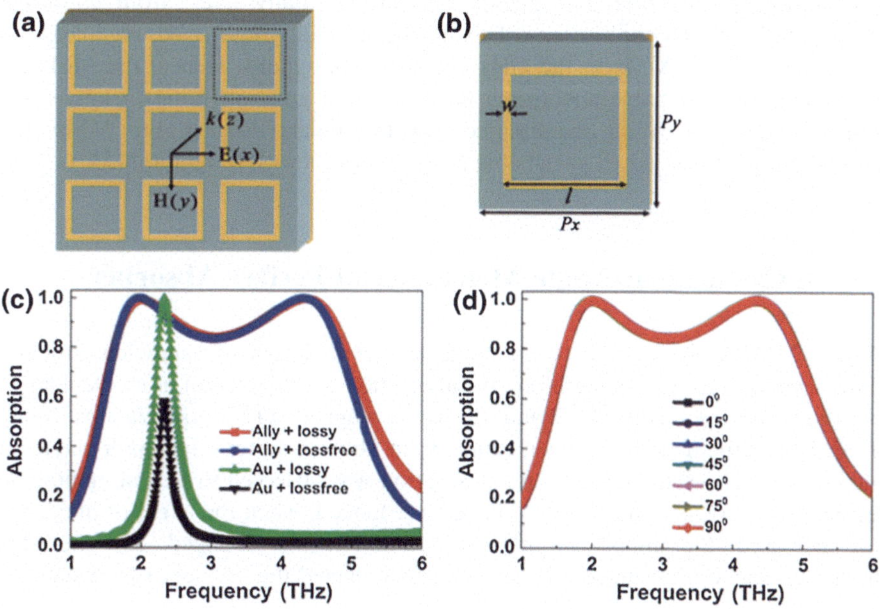

Fig. 6.10 **a** Structural schematic of the proposed low-conductivity alloy MM absorber: the *black dotted line* represents a unit cell. **b** Dielectric layer (thickness t = 13 μm) of a dielectric constant of 3 (1 + 10.06) and alloy (Al–Si–Mg) thickness of 0.4 μm with a conductivity of $\sigma = 4.09 \times 10^5$ S/m. The optimized parameters are $P = Px = Py = 30$, l = 22 and w = 1.0 μm. **c** Simulated absorption spectra of proposed alloy and Au absorbers with two different loss conditions (loss free and lossy). **d** Dependence of the absorption spectra on the polarization angle of incident wave for the proposed alloy absorber [14]

broadband absorption also reached 118.2 %. By using a high conductivity MM absorber (Au with a conductivity of $\sigma = 4.09 \times 10^7$ S/m), only a single absorption peak was observed at 2.33 THz (f_3), with absorption of 99.67 %. At the same time, the bandwidth of proposed alloy absorber was 11.3 times larger than that of the high-conductivity (Au) absorber. According to the *LC*-circuit model, the absorption frequency of the absorber could be expressed by

$$f = \frac{1}{2\pi\sqrt{LC/2}} \sim \frac{1}{l}. \tag{6.10}$$

From (6.10), the absorption frequency was inversely proportional to the length of the square ring. Consequently, with the other geometric parameters fixed, the bandwidth of absorption spectrum was controlled by changing length l of patterned metallic layer. In particular, the broadband absorption turned out to be polarization-insensitive with respect to the normally-incident EM wave (as shown in Fig. 6.10d). This was qualified owing to the high symmetry of square-ring structure (four-fold symmetric structure).

The control of MM structure, with proper adjustment of the absorption according to EM-wave polarization, has greatly affected our daily lives, from consumer products to high-tech applications. This has been rapidly developed, together with the development of MMs in the GHz and the THz regions. Hence, the MMs for controllable or switchable absorption have attracted great attention in optics and photonics research. With manipulation of the polarization in the MMs, the embedded diode has led to switchable absorption in the GHz region [15].

6.3 Wide-Incident-Angle Metamaterial Perfect Absorber

Mostly, MMPAs have been based upon the novel design of sandwich structure (three-layer design) [2–4], therefore, the absorption was a function of incident angle for both transverse-electric (TE) and transverse-magnetic (TM) polarizations. Many of MMPA structures have been studied on the absorption for high angle at microwave, THz or IR. Usually, the absorption was observed to have a monotonic decrease by increasing the angle of incidence. Indeed, when the angle of incidence was increased, the impedance-matching condition was changed, and the MM structure became impedance-mismatched and reflected the incident EM wave.

(A) Conventional Approach

Right after the first exploration of MMPA [2], the first computational and experimental results on the wide-angle MMPA have been acquired by Tao et al. [16]. The absorption at THz remained above 99 % at the TM polarization for incident angle less than 80°, and at the TE polarization the absorption reached above 90 % for incident angle up to 50°.

The MM absorber consisted of two metallic layers separated by a dielectric spacer. An array of split-ring resonators (SRRs) was placed at the top layer which was mainly responsible for tuning $\varepsilon(\omega)$, and the bottom layer of metallic layer drove circulating currents between two metallic layers. The top layer consisted of an array of 200-nm-thick Au with the parameters of side length $b = 25.9$, capacitor length $c = 10.8$, capacitor gap $g = 1.4$, and line width $w = 3\ \mu m$ (shown in Fig. 6.11a). The unit cell a was 36 μm, and two dielectric layers t_1 and t_2 were with thickness of 8 μm. In the interested frequency range, the dielectric layer (Fig. 6.11b) was modeled as a polyimide with frequency-independent permittivity $\varepsilon = 2.88 + 0.09i$. With the highly flexible polyimide substrate (Fig. 6.11c), the MMPA structure was able to use in non-planar applications as it could easily cover around the objects with diameter larger than 6 mm.

According to the optimized parameters in Fig. 6.11, a single absorption peak was observed with the absorption of 99.9 % at 1.6 THz. For the TE polarization, even though the absorption was reduced with increasing the angle of incidence, but still remained larger than 89 % below the angle of incidence of 50° (Fig. 6.12a and c). Besides, a slight frequency shift of ~30 GHz was also noted from 0 to 80°. In case of the TM polarization, interestingly, the results showed that the absorption remained greater than 99 % for incident angle from 0 even to 80° (Fig. 6.12b and d).

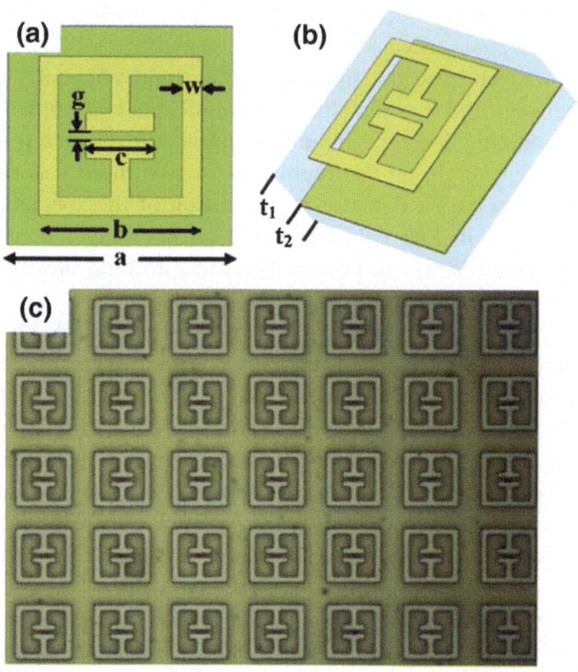

Fig. 6.11 THz MM absorber consisting of two metallic layers and two dielectric layers. **a** Front view and **b** perspective view of the absorber. **c** Photo of the fabricated sample [16]

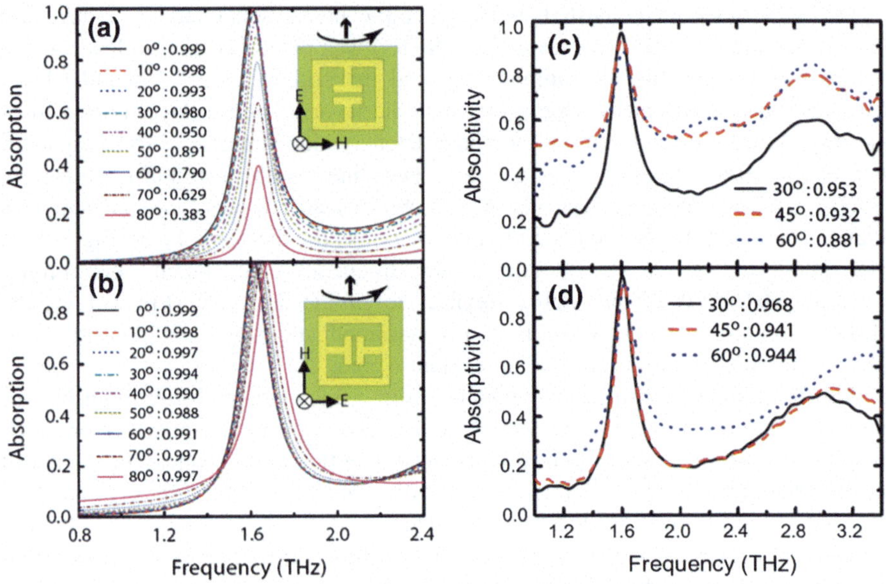

Fig. 6.12 Simulated absorption spectra according to incident angle for **a** TE and **b** TM incident radiation. Experimental absorption spectra according to incident angle for **c** TE and **d** TM incident radiation [16]

It was because the magnetic field could efficiently drive the circulating currents between the two metallic parts at all incident angles. Many studies have shown a monotonic decrease in the absorption at resonance for the TE mode as a function of incident angle up to roughly 40°, while there was only slight change for the TM mode [17–19]. Because of the reason of the special structural design, the MMPA structure in Fig. 6.11 depended on not only the magnetic response but also the in-plane electric response, therefore, it could absorb the EM wave up to a larger angle of incidence even for the TE mode.

The structure configuration of some chiral MMs could also be incident-angle independent [20, 21]. Figure 6.13 presents a sample of chiral MM. The unit cell was composed of two identical SRRs separated by a dielectric substrate and interconnected by bias (Fig. 6.13a). Chiral MM was firstly introduced to retrieve negative refraction [22], then by adding a ground of copper plate and covered with a dielectric plate, a chiral MM absorber was made (Fig. 6.13b). The structure design of chiral MM absorber was to somehow transform the reflected wave for the case of oblique incident angle. However, the polarization-transformed reflected spectrum was less than 3 % for incident angle up to 80°. The absorption was calculated correctly by $A = 1 - R_{EE} - R_{HE}$ for TE radiation, and $A = 1 - R_{HH} - R_{EH}$ for TM radiation, where subscripts E and H refer to the TE and the TM polarization, respectively. Therefore, R_{HE} and R_{EH} indicate the polarization transformed due to the reflection. The absorption spectra are presented in Fig. 6.13c–f. For TE

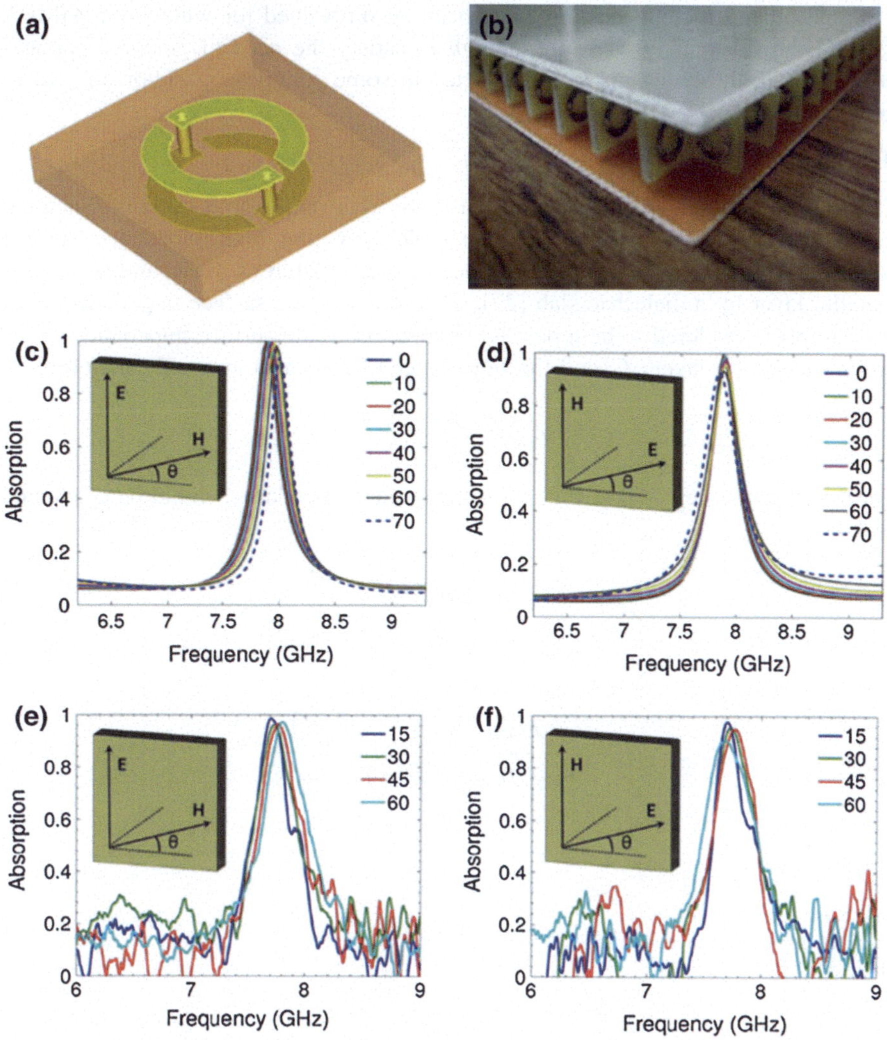

Fig. 6.13 **a** Structure of the chiral SRR. **b** Metal ground plate and cover plate were attached to the MM slab to obtain a MM absorber. The metal ground plate was also a PCB with copper on one side. The cover plate was a PCB with no copper cover. The absorption results at different angles for **c** simulation and **e** experiment for the TE mode, together with **d** simulation and **f** experiment for the TM mode [20]

polarization, nearly unity absorption was obtained at normal incidence and remained above 98 % for, at least, 60° and above 90 % until 70° of incident angle. In case of TM polarization, the absorption was above 90 % for nearly all incident angles.

Some other structural designs have been accomplished for wide-angle MMPA [23–26], however, they were not enough to satisfy the need of practical applications. Indeed, these designs were restricted in some parameters, and/or difficult to fabricate, etc.

(B) High-impedance Surface

The analyze on the dependence of the MM absorber on the polarization of incident radiation, have been needed. In the simple way, the high-impedance surface structure is illustrated in Fig. 6.14, the square structure was separated from a metallic layer by a dielectric slab [27]. Theoretically, the surface impedance, Z_{inp}, was simply considered to be a parallel connection of the surface impedance of the ground dielectric layer, Z_s, and the impedance of the patch array, Z_p, therefore

$$Z_{inp}^{-1} = Z_p^{-1} + Z_s^{-1}.$$

For the structure depicted in Fig. 6.14, the surface impedances could be calculated as follow.

$$Z_{inp}^{TE} = \frac{j\omega\mu_0 \frac{\tan(\beta h)}{\beta}}{1 - 2k_{eff}\alpha \frac{\tan(\beta h)}{\beta}\left(1 - \frac{\sin^2(\theta)}{\varepsilon_r + 1}\right)}, \tag{6.11}$$

$$Z_{inp}^{TM} = \frac{j\omega\mu_0 \frac{\tan(\beta h)}{\beta}\left(1 - \frac{\sin^2(\theta)}{\varepsilon_r}\right)}{1 - 2k_{eff}\alpha \frac{\tan(\beta h)}{\beta}\left(1 - \frac{\sin^2(\theta)}{\varepsilon_r}\right)}, \tag{6.12}$$

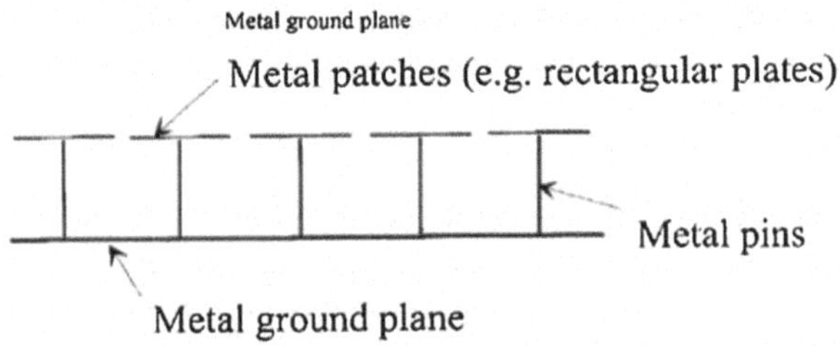

Fig. 6.14 Geometry of the artificial high-impedance surface [27]

where $\beta = \sqrt{k_0^2 \varepsilon_r - k_t^2}$ is the normal component of the wave vector travelling in the dielectric substrate, h is the thickness of dielectric substrate, $k_{eff} = k_0 \sqrt{\varepsilon_{eff}}$ is the effective wave vector, $\varepsilon_{eff} = (\varepsilon_r + 1)/2$ is the effective permittivity, ε_r is the permittivity of substrate, θ is the incident angle and $\alpha = \frac{k_{eff}}{D} \ln\left(\frac{1}{\sin(\pi\omega/2D)}\right)$ is grid parameter, D is the unit-cell size, and ω is the spacing between two metallic square.

Usually, the thickness of substrate is much smaller than the wavelength, therefore, $\tan(\beta h)/\beta \approx h$. Equations (6.11) and (6.12) become:

$$Z_{inp}^{TE} = \frac{j\omega\mu_0 h}{1 - 2k_{eff}\alpha h\left(1 - \frac{\sin^2(\theta)}{\varepsilon_r + 1}\right)}, \tag{6.13}$$

$$Z_{inp}^{TM} = \frac{j\omega\mu_0 h\left(1 - \frac{\sin^2(\theta)}{\varepsilon_r}\right)}{1 - 2k_{eff}\alpha h\left(1 - \frac{\sin^2(\theta)}{\varepsilon_r}\right)}. \tag{6.14}$$

From (6.13) and (6.14), one can see that the angle-dependent terms have the permittivity of substrate, ε_r, to appear in the denominator. To diminish the effect of incidence angle on the surface impedance, in other words, to make the absorber work with higher incident angle, we can increase the permittivity of substrate. For example, when $\varepsilon_r = 9(1 + i0.2)$, the absorption of structure, which was shown in Fig. 6.14, was higher than 90 % for TM polarization even up to the incident angle of 60°. The results are shown in Fig. 6.15 [27].

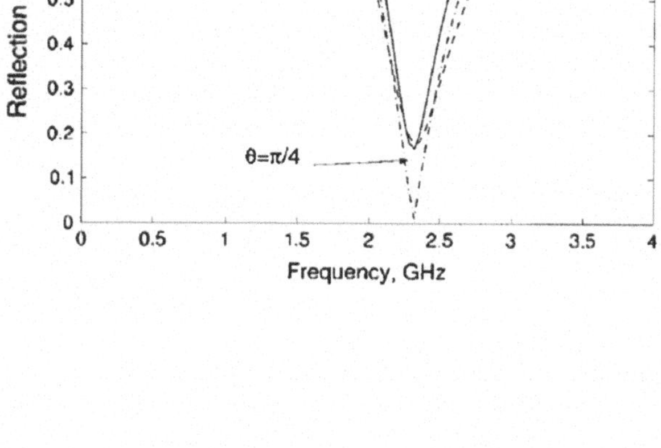

Fig. 6.15 Power reflection factors for the incidence angles of square structure in TM mode. The parameters of the absorber are $D = 10$, $w = 0.1$, and $h = 3$ mm [27]

C. Perfectly-matched-layer-like Metamaterial Perfect Absorber Structure

The idea for an artificial absorbing layer regardless of frequency and/or incident angle, was introduced mathematically by Berenger in 1994 [28], called PML. The key idea of PML was a medium that did not reflect at the interface and then strongly absorbed from the interior region. Later, another formulation of PML has been introduced and became more popular owing to its simplicity and efficiency, called uniaxial PML or UPML. An UPML is a PML layer with an artificial anisotropic absorbing material. Taking advantage of UPML, we can computationally introduce an angle-independent absorber material.

The Fresnel equations are as follows.

$$\begin{aligned} S_{11}^{TE} &= \frac{Z_2 \cos\theta_i - Z_1 \cos\theta_t}{Z_2 \cos\theta_i + Z_1 \cos\theta_t}, \\ S_{21}^{TE} &= \frac{2\, Z_2 \cos\theta_i}{Z_2 \cos\theta_i + Z_1 \cos\theta_t}. \end{aligned} \tag{6.15}$$

$$\begin{aligned} S_{11}^{TM} &= \frac{Z_2 \cos\theta_t - Z_1 \cos\theta_i}{Z_2 \cos\theta_t + Z_1 \cos\theta_i}, \\ S_{21}^{TE} &= \frac{2\, Z_2 \cos\theta_i}{Z_2 \cos\theta_t + Z_1 \cos\theta_i}. \end{aligned} \tag{6.16}$$

We now temporary dismiss the presence of transmission in (6.15) and (6.16), owing to the fact that the transmission can be entirely absorbed when we put the dissipation part in both permittivity and permeability. The next work is to prevent the reflection, and we need the impedance matching for any incident angle or polarization.

To satisfy this condition, the reflections in both (6.15) and (6.16) must vanish, in other words, $Z_2 = Z_1 \frac{\cos\theta_t}{\cos\theta_i}$ and $Z_2 = Z_1 \frac{\cos\theta_i}{\cos\theta_t}$. Figure 6.16 shows the reflection and the transmission of EM waves at an interface. Unfortunately, we have both impedances and angles in this term, so we cannot assure that the impedance is matched for all of the angles. To overcome this problem, Sacks et al. [29] introduced an anisotropic material, in which the permittivity and the permeability were in form of a tensor. The impedance of PML Z is $\sqrt{\frac{\mu_r}{\varepsilon_r}}$, and the work is to find a reflectionless layer or Z = 1. They chose $[s] = [\mu_r] = [\varepsilon_r] = \begin{bmatrix} a & 0 & 0 \\ 0 & b & 0 \\ 0 & 0 & c \end{bmatrix}$. According to the magnitude and phase matching at the boundary of two surfaces (the mathematical transformation is fully described in [28]), the Snell's law become

$$\sqrt{bc}\ \sin\theta_t = \sin\theta_i, \tag{6.17}$$

and the reflection is represented by

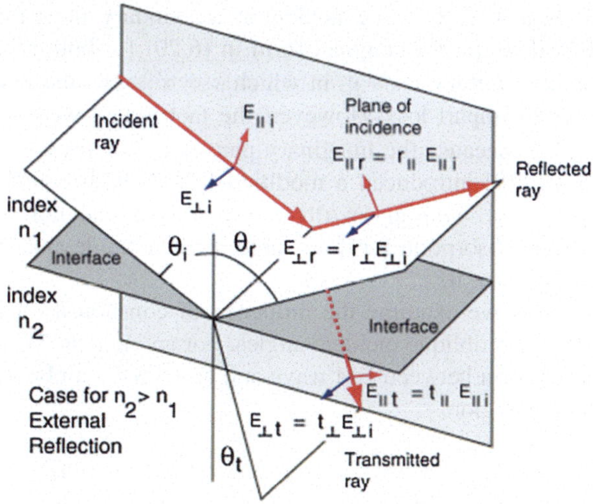

Fig. 6.16 Reflection and transmission of EM waves at an interface [Origin: http://hyperphysics.phy-astr.gsu.edu/hbase/phyopt/reflin.html]

$$S_{11}^{TE} = \frac{\cos\theta_i - \sqrt{\frac{b}{a}}\cos\theta_t}{\cos\theta_i + \sqrt{\frac{b}{a}}\cos\theta_t}, \tag{6.18}$$

$$S_{11}^{TM} = \frac{\sqrt{\frac{b}{a}}\cos\theta_t - \cos\theta_i}{\cos\theta_i + \sqrt{\frac{b}{a}}\cos\theta_t}. \tag{6.19}$$

We can freely to choose $bc = 1$, to simplify the phase matching in (6.17), leading to $\theta_t = \theta_i$, and the reflection coefficients are no longer a function of incident angle. By also postulating $a = b$, the interface comes to be perfectly reflectionless for any frequency, angle of incident, and polarization. Thus, we can write our UPML in terms of the constitution parameters

$$[s] = [\mu_r] = [\varepsilon_r] = \begin{bmatrix} s & 0 & 0 \\ 0 & s & 0 \\ 0 & 0 & s^{-1} \end{bmatrix}. \tag{6.20}$$

Equation (6.20) represents a PML layer that works for the wave travelling in the +z direction, for the 3-dimensional PML and incorporating loss, the PML layer now has the parameters as follows.

$$[s] = [s_x]\cdot[s_y]\cdot[s_z] = \begin{bmatrix} \frac{s_y s_z}{s_x} & 0 & 0 \\ 0 & \frac{s_x s_z}{s_y} & 0 \\ 0 & 0 & \frac{s_y s_x}{s_z} \end{bmatrix}, \tag{6.21}$$

where $s_r = \alpha_r' + i\alpha_r''$ and $r = (x, y, z)$.

In reality, for wave incident on a boundary along the x–y plane, we only need the UPML to take a diagonal form in (6.20) for both relative permittivity and permeability, $dia[g(s,s,s^{-1})]$, in which $s = \alpha' + i\alpha''$, the imaginary part α'' must be positive to impart loss. However, the fact is that we experimentally cannot create an UPML because the imaginary part of s^{-1} is always negative. Due to this fact, Ye et al. [30] introduced a modified UPML with constitutive tensors in the form of $[\mu_r] = [\varepsilon_r] = dia[g(s,s,h)]$, where h is a real number. This can retrieve nearly perfect absorption for quite large incident angle and, most importantly, we are able to fabricate it.

Now we examine the influence of constant h on the absorption of a modified UPML at oblique incident angles. For an incident TE wave, the field components in interaction between EM wave and an UPML can be described according to [31]. In the air region,

$$E = E_y\hat{y} = \left(e^{i(k_x x + k_z z)} + Ae^{i(k_x x - k_z z)}\right)\hat{y}, \tag{6.22}$$

$$\begin{aligned} H &= \nabla \times \frac{E}{i\omega\mu_0} = H_x\hat{x} + H_z\hat{z} \\ &= \left(-\frac{k_z}{\omega\mu_0}e^{i(k_x x + k_z z)} + \frac{Ak_z}{\omega\mu_0}e^{i(k_x x - k_z z)}\right)\hat{x} \\ &\quad + \left(-\frac{k_x}{\omega\mu_0}e^{i(k_x x + k_z z)} + \frac{Ak_x}{\omega\mu_0}e^{i(k_x x - k_z z)}\right)\hat{z}, \end{aligned} \tag{6.23}$$

where $k_x = k_0\sin\theta$, $k_z = k_0\cos\theta$ and $A = S_{11}$. In modified UPML region,

$$E_1 = E_{1y}\hat{y} = \left(Be^{i(k_x x + k_z z)} + Ce^{i(k_x x - k_z z)}\right)\hat{y}, \tag{6.24}$$

$$\begin{aligned} H_1 &= \nabla \times \frac{E_1}{i\omega\mu_0} = H_{1x}\hat{x} + H_{1z}\hat{z} \\ &= \left(-\frac{Bk_{1z}}{\omega\mu_0\mu_{rx}}e^{i(k_{1x}x + k_{1z}z)} + \frac{Ck_{1z}}{\omega\mu_0\mu_{rx}}e^{i(k_{1x}x - k_{1z}z)}\right)\hat{x} \\ &\quad + \left(-\frac{Bk_{1x}}{\omega\mu_0\mu_{rz}}e^{i(k_{1x}x + k_{1z}z)} + \frac{Ck_{1x}}{\omega\mu_0\mu_{rz}}e^{i(k_{1x}x - k_{1z}z)}\right)\hat{z}, \end{aligned} \tag{6.25}$$

where $k_{1x}^2/\mu_{rz} + k_{1z}^2/\mu_{rx} = \varepsilon_{ry}k_0^2$. B and C are constants. Boundary conditions say that $k_x = k_{1x}, E_y = E_{1y}$ and $H_x = H_{1x}$ at z = 0, and $E_{1y} = 0$ when z has the value of thickness of the UPML. The reflection can finally be obtained to be

$$S_{11} = \frac{-k_z\cos(k_z L) - \mu_{rx}k_0\cos\theta\sin(k_z L)i}{k_z\cos(k_z L) - \mu_{rx}k_0\cos\theta\sin(k_z L)i}e^{-2ik_0\cos\theta L}, \tag{6.26}$$

where $k_{1z} = k_0\sqrt{\varepsilon_{ry}\mu_{rx} - \mu_{rx}\sin^2\theta/\mu_{rz}}$ and $\varepsilon_{ry} = \mu_{rx} = \alpha$, $\mu_{rz} = h$. It is clearly seen that the real constant h affects only wave vector k_1. To check the influence of h on

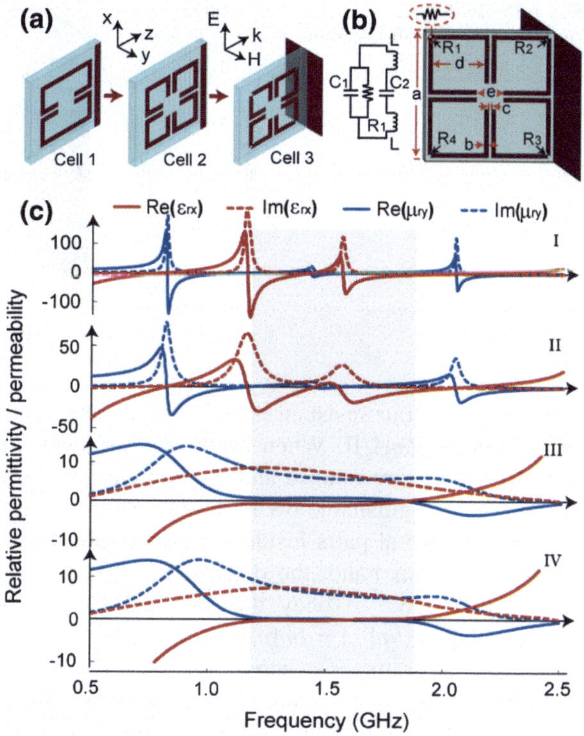

Fig. 6.17 Design of the unit cell of an ultra-wideband MMPA. **a** Introduction of multiple resonances in SRR and rod cell. **b** Topology of the new cell and related equivalent circuit model. **c** Retrieved constitutive parameters for different resistances [32]

the absorption under the condition that the UPML had large imaginary $\alpha(\alpha = -0.001 + 14.7i)$, the absorption as a function of the incident angle for several h values from 0.1 to 10 were calculated.

Ye et al. reported the modified PML MMPA [32]. Figure 6.17 shows design of the unit cell of the ultra-wideband MM absorber. They started the basic sub-wavelength unit cell consisting of metallic split-ring resonator (SRR) and rod (Cell 1 in Fig. 6.17a). In order to realize the multiple electric and magnetic resonances without increasing the size of cell, they split the original SRRs into four smaller, centrosymmetric SRRs, as illustrated by "Cell 2." To introduce independently tunable loss for each SRR, the outside corner of each SRR is cut and a lumped resistor is soldered (Cell 3). The final topology of the new cell is shown in Fig. 6.17b. In each cell, four SRRs are printed on a 1.5-mm-thick FR4 substrate with a dielectric constant of 4.4 and a loss tangent of 0.02. To ensure commercial surface-mount resistors exhibiting negligible reactance, the dimensions of the unit cell are chosen to be a = 40, b = 1.2, c = 0.4, d = 16.4, and e = 8 mm, such that multiple resonances appear in a band centered at 1.5 GHz. Around this frequency, the periodicity of the unit cells is $\lambda/5$, and the MMPA can be treated as an effective medium to retrieve effective constitutive parameters using a homogenization algorithm. As shown in panel I of Fig. 6.17c, three magnetic resonances and two electric resonances appeared in the band from 500 MHz to 2.5 GHz. After

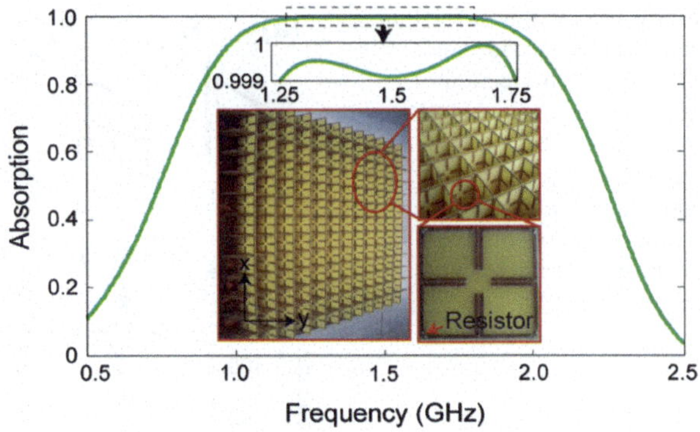

Fig. 6.18 Full-wave simulation of the fabricated MMPA sample [32]

increasing the four resistances to 10 Ω, all the bandwidths of the resonances are broadened in panel II. When the four resistances are increased to be 100 Ω, the second magnetic resonance and the two electric resonances inside the grey region become indistinguishable owing to the increased dissipation in panel III, and both imaginary and real parts inside the grey region become flatter. Finally, to achieve nearly PA in this band, the dispersions of both complex permittivity ε_r and permeability μ_r have to obey the modified PML model in the whole band, i.e., $Re[\varepsilon_r] = Re[\mu_r]$, $Im[\varepsilon_r] = Im[\mu_r]$, and $Im[\varepsilon_r]$ and $Im[\mu_r]$ should be larger than 2. By performing a parameter sweep of the four resistances, the optimized set of 203, 122, 39 and 63 Ω were for resistors 1–4, respectively. With these values, the dispersions of the complex permittivity and permeability are nearly equal in the frequency band from 1.2 to 1.8 GHz, as shown in panel IV. In this band, the real parts of permittivity and permeability were close to zero, while the imaginary parts were all larger than 5.

The implemented $520 \times 520\ \text{mm}^2$ sized sample includes 312 unit cells and 1248 surface-mount resistors, as shown in the inset of Fig. 6.18. Due to the cross coupling induced by the interlock, resistors $R1$, $R2$, $R3$ and $R4$ have been tuned to be 226, 324, 280 and 63.5 Ω, respectively. The simulated absorption was given in Fig. 6.18. With these modified resistances, the bandwidth for 99.9 % of absorption is now 0.47 GHz (from 1.27 to 1.75 GHz), corresponding to a relative bandwidth of 31 %. The center of the sample was carefully aligned with the center of the antenna aperture, as shown in the inset of Fig. 6.18a. The normalized absorption for both y- and x-polarized incidences are calculated by $1 - |S_{11}|^2$, and are compared with the simulation results in Fig. 6.19a. For the x-polarized incidence, the bandwidth in which the absorption is larger than 99 % is 0.49 GHz (from 1.3 to 1.79 GHz), corresponding to a relative bandwidth of 31.5 %. For the y-polarized incidence (red line), the power absorption is similar to the x polarization. Figure 6.19b plots the normalized reflection versus incident angles. When the incident angles were smaller than 30°, the absorption below −20 dB still occurred over an ultrawide bandwidth of 0.59 GHz (from 1.22 to 1.81 GHz), corresponding to a relative bandwidth of 39 %. Base on high polarization sensitivity, certain MMPA might have potential to be used as polarization detectors, sensors, polarizers and so on [33].

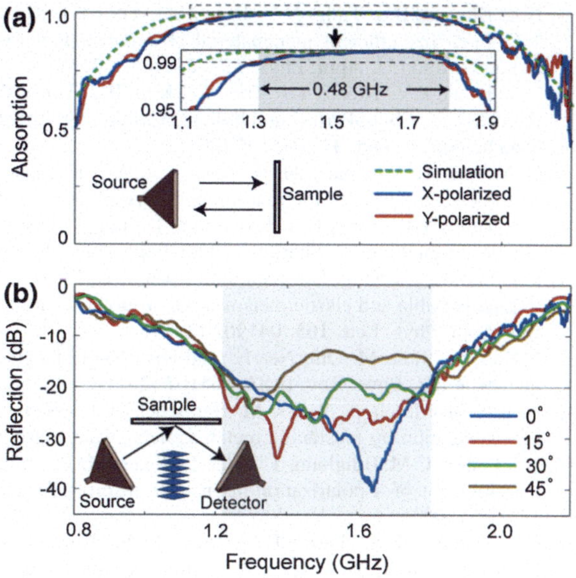

Fig. 6.19 Experimental absorption of the fabricated MMPA sample. **a** Normal incidence. **b** Oblique incidences [32]

6.4 Summary

In this chapter, we have reviewed the dependence of MMPA on the polarization and the incident angle of EM wave. Usually, MMPA is usually polarization-dependent, however, by exploiting the symmetry of the front metallic layer, we can create polarization-independent MMPAs. Studies for wide-incident-angle MMPAs have shown that the absorption was decreased when the incident angle was increased, however, we need the MMPA to work for wide incident angle for the practical applications. For example, the absorption should exceed 90 % simultaneously for incident angle up to, at least, 50°. Some special structure can satisfy the requirements, especially, the theory shows that, as higher permittivity value of the dielectric layer, the MMPA can work for wider incident angle.

To be in more detail, the theory of UPML was introduced as a solution of incident-angle-independent MMPA, however, we cannot make a truly UPML in real world. Therefore, a modified UPML layer was pursued by removing the unavailable parameters in the conventional UPML. The PML-like structure has been proved to be wide-incident-angle MMPA.

References

1. M. Planck, Uber das Gestz der energieverteilung im normalspektrum, Ann. Phys. (Leipzig) **309**, 553 (1901)
2. N.I. Landy, S. Sajuyigbe, J.J. Mock, D.R. Smith, W.J. Padilla, Perfect metamaterial absorber. Phys. Rev. Lett. **100**, 207402 (2008)

3. B.X. Khuyen, B.S. Tung, N.V. Dung, Y.J. Yoo, Y.J. Kim, K.W. Kim, V.D. Lam, J.G. Yang, Y.P. Lee, Size-efficient metamaterial absorber at low frequencies: design, fabrication, and characterization. J. Appl. Phys. **117**, 243105 (2015)
4. N.V. Dung, P.V. Tuong, Y.J. Yoo, Y.J. Kim, B.S. Tung, V.D. Lam, J.Y. Rhee, K.W. Kim, Y. H. Kim, L.Y. Chen, Perfect and broad absorption by the active control of electric resonance in metamaterial. J. Opt. **17**, 045105 (2015)
5. Q.-Y. Wen, H.-W. Zhang, Q.-H. Yang, Y.-S. Xie, K. Chen, Y.-L. Liu, Terahertz metamaterials with VO_2 cut-wires for thermal tunability. Appl. Phys. Lett. **97**, 1111 (2010)
6. B.S. Tung, N.V. Dung, B.X. Khuyen, N.T. Tung, P. Lievens, Y.P. Lee, V.D. Lam, Thermally tunable magnetic metamaterials at THz frequencies. J. Opt. **15**, 075101 (2013)
7. Y.J. Yoo, H.Y. Zheng, Y.J. Kim, J.Y. Rhee, J.-H. Kang, K.W. Kim, H. Cheong, Y.H. Kim, Y. P. Lee, Flexible and elastic metamaterial absorber for low frequency, based on small-size unit cell. Appl. Phys. Lett. **105**, 041902 (2014)
8. J. Hao, L. Zhou, M. Qiu, Nearly total absorption of light and heat generation by plasmonic metamaterials. Phys. Rev. B **83**, 165107 (2011)
9. N. Liu, M. Mesch, T. Weiss, M. Hentschel, H. Giessen, Nearly total absorption of light and heat generation by plasmonic metamaterials. Nano Lett. **10**, 2342 (2010)
10. N.I. Landy, C.M. Bingham, T. Tyler, N. Jokerst, D.R. Smith, W.J. Padilla, Design, theory, and measurement of a polarization-insensitive absorber for terahertz imaging. Phys. Rev. B **79**, 125104 (2009)
11. P.V. Tuong, J.W. Park, J.Y. Rhee, K.W. Kim, W.H. Jang, H. Cheong, Y.P. Lee, Polarization-insensitive and polarization-controlled dual-band absorption in metamaterials. Appl. Phys. Lett. **102**, 081122 (2013)
12. S. Ghosh, S. Bhattacharyya, Y. Kaiprath, K.V. Srivastava, Bandwidth-enhanced polarization-insensitive microwave metamaterial absorber and its equivalent circuit model. J. Appl. Phys. **115**, 104503 (2014)
13. S. Bhattacharyya, K.V. Srivastava, Triple band polarization-independent ultra-thin metamaterial absorber using electric field-driven LC resonator. J. Appl. Phys. **115**, 064508 (2014)
14. B.-X. Wang, L.-L. Wang, G.-Z. Wang, W.-Q. Huang, X.-F. Li, X. Zhai, A simple design of a broadband, polarization-insensitive, and low-conductivity alloy metamaterial absorber. Appl. Phys. Express **7**, 082601 (2014)
15. B. Zhu, Y. Feng, J. Zhao, C. Huang, T. Jiang, Switchable metamaterial reflector/absorber for different polarized electromagnetic waves. Appl. Phys. Lett. **97**, 051906 (2010)
16. H. Tao, C.M. Bingham, A.C. Strikwerda, D. Pilon, D. Shrekenhamer, N.I. Landy, K. Fan, X. Zhang, W.J. Padilla, R.D. Averitt, Highly flexible wide angle of incidence terahertz metamaterial absorber: design, fabrication, and characterization. Phys. Rev. B **78**, 241103 (2008)
17. Z. Bo, W. Zheng-Bin, Y. Zhen-Zhong, Z. Qi, Z. Jun-Ming, F. Yi-Jun, J. Tian, Planar metamaterial microwave absorber for all wave polarizations. Chin. Phys. Lett. **26**, 114102 (2009)
18. W. Zhu, X. Zhao, B. Gong, L. Liu, B. Su, Optical metamaterial absorber based on leaf-shaped cells. Appl. Phys. A: Mater. Sci. **102**, 147 (2011)
19. C.M. Watts, X. Liu, W. J. Padilla, Metamaterial electromagnetic wave absorbers. Adv. Mater. **24**, OP98 (2012)
20. B. Wang, T. Koschny, C.M. Soukoulis, Wide-angle and polarization-independent chiral metamaterial absorber. Phys. Rev. B **80**, 033108 (2009)
21. Y. Li, Q. Huang, D.C. Wang, X. Li, M.H. Hong, X.G. Luo, Polarization-independent broadband terahertz chiral metamaterials on flexible substrate. Appl. Phys. A Mater. Sci. Process. **115**, 57 (2014)
22. S. Zhang, Y.-S. Park, J. Li, X. Lu, W. Zhang, X. Zhang, Negative refractive index in chiral metamaterials. Phys. Rev. Lett. **102**, 023901 (2009)
23. B. Zhu, Z. Wang, C. Huang, Y. Feng, J. Zhao, T. Jiang, Polarization insensitive metamaterial absorber with wide incident angle. Prog. Electromagn. Res. **101**, 231 (2010)

24. B.-X. Wang, L.-L. Wang, G.-Z. Wang, W.-Q. Huang, X. Zhai, Broadband, polarization-insensitive and wide-angle terahertz metamaterial absorber. Phys. Scripta **89**, 115501 (2014)
25. D. Chaurasiya, S. Ghosh, S. Bhattacharyya, K.V. Srivastava, An ultrathin quad-band polarization-insensitive wide-angle metamaterial absorber. Microw. Opt. Technol. Lett. **57**, 697 (2015)
26. X. Li, H. Liu, Q. Sun, N. Huang, Photo. Nano. Fund. Appl. (2015)
27. S.A. Tretyakov, S.I. Maslovski, Thin absorbing structure for all incidence angles based on the use of a high-impedance surface. Microw. Opt. Technol. Lett. **38**, 175 (2003)
28. J.-P. Berenger, A perfectly matched layer for the absorption of electromagnetic waves. J. Comput. Phys. **114**, 185 (1994)
29. Z.S. Sacks, D.M. Kingsland, R. Lee, J.-F. Lee, A perfectly matched anisotropic absorber for use as an absorbing boundary condition. IEEE Trans. Antenna. Propag. **43**, 1460 (1995)
30. D. Ye, Z. Wang, K. Xu, B. Zhang, J. Huangfu, C. Li, L. Ran, Towards experimental perfectly-matched layers with ultra-thin metamaterial surfaces. IEEE Trans. Antenna. Propag. **60**, 5164 (2012)
31. J.A. Kong, *Theory of Electromagnetic Waves* (Wiley-Interscience, New York, 1975)
32. D. Ye, Z. Wang, K. Xu, H. Li, J. Huangfu, Z. Wang, L. Ran, Ultrawideband dispersion control of a metamaterial surface for perfectly-matched-layer-like absorption. Phys. Rev. Lett. **111**, 187402 (2013)
33. Y.J. Kim, J.M. Kim, Y.J. Yoo, P.V. Tuong, H. Zheng, J.Y. Rhee, Y.P. Lee, Dual-absorption metamaterial controlled by electromagnetic polarization. J. Opt. Soc. Am. B **31**, 2744 (2014)

Chapter 7
Perspectives

Abstract With simple design and geometrical scalability, the investigated metamaterial (MM) absorbers might operate at higher or lower frequency with perfect absorption and be modified to reveal other capabilities relevant to the practical applications. The absorbers for GHz and MHz ranges are being developed to cover wider spectral regime, and to be independent of polarization and the incident angle of incident EM wave, which can be used in suppressing the radiation from mobile and other electric equipments. For MMs operating in the GHz and the lower-frequency ranges, another important challenging aspect should be overcome: the relatively (to the working wavelength) smaller size of the unit cell. The flexible MMPAs made by simpler fabrication processes come to be achieved. The present MMPAs also need simpler design, easier fabrication, lower production cost, and higher fabrication tolerance. Multi-band MMPAs using single pattern are interesting for the simple fabrication and the relevant advantages. Stealth science and technology based on microwave (L, X and K_u bands) MMPAs are going to be further developed. We feel confident that the MM revolution in the field of absorption has just started, and will present exciting new developments in the realm of practical applications in the coming years by overpassing the entry barrier and replacing the conventional applications using existing materials/structures, and by exploiting new ones. The industrial applications of MMs will be accelerated, which is the transition of MMs from basic and applied research into real world applications.

With simple design and geometrical scalability, the investigated metamaterial (MM) absorbers might operate at higher or lower frequency with perfect absorption and be modified to reveal other capabilities relevant to the practical applications [1]. The works on dual-band MM absorber promises for further applications, especially, to sub-wavelength detectors or filters [2]. The multi-plasmonic perfect absorbers (PAs) are also achieved in THz, mid-infrared, and visible regions of EM wave. This kind of investigations guarantees advanced MMPAs and their applications at higher frequencies [3]. The absorbers for GHz and MHz ranges are being developed to cover wider spectral regime, and to be independent of polarization and the incident angle of incident EM wave, which can be used in suppressing the radiation from

Y.P. Lee et al., *Metamaterials for Perfect Absorption*,
Springer Series in Materials Science 236, DOI 10.1007/978-981-10-0105-5_7

mobile and other electric equipments [4]. They include flexible printed circuit and camera unit in cellular phone, interface and data driver in PDP television, interface in large LCD panel, and GPS unit and signal input in car navigation [5]. They also comprise infrastructures relevant to EM wave, such as radar mast of a ship, building wall in front of a broadcasting station, and ceiling and wall of intelligent-transportation-system toll gate. Especially, for MMs operating in the GHz and the lower-frequency ranges, another important challenging aspect should be overcome: the size of the unit cell. The operating frequency, or equivalently wavelength, is critically determined by the lattice parameter. Since the lattice parameter increases for MMPAs operating in the low-frequency range and, as a result, the same is true for the total size of sample, deep-subwavelength (lattice parameter $\leq \lambda/10$) MMPAs should be investigated for immediate and versatile applications. The flexible MMPAs for GHz, by employing ink-jet printing, are pursued recently. On the other hand, inks of more kinds of electrical conductivity should be prepared to realize high-performance flexible MMPAs, and the reproducibility of the process also matters. The flexibility continues to be the issue for other frequency ranges, such as THz and visible, of EM wave, and furthermore, the flexible MMPAs made by simpler fabrication processes come to be achieved.

The optimized MMPAs in the THz region could work as improved detectors for imaging and detecting the harmful elements to human, and for detecting heat in enhanced and spectrally-selective way. The match of MMPAs in the THz region makes the relevant important applications of special interest worldwide, which comprise non-destructive quality and process control, and environmental monitoring implementations [6]. On some of these devices, more sophisticated characteristics such as tunability, multi-band or broadband spectral response, and incident-angle- and/or polarization-independent responses, should be demonstrated. These are also necessary for higher frequency (infrared and visible ranges) MMPAs, leading to advanced applications in optics and photonics.

Plasmonic MM with nearly PA of EM wave in visible frequency range keeps being studied [7]. The absorption mechanism should be further elucidated. Only the interference theory cannot explain the broadband perfect absorption, and in parallel, plasmonic coupling and multi-excitation of plasmon resonance due to the reflective nature of copper-base film, also contribute to the high absorption of the structure. Owing to the simple fabrication technique, the production cost is very low compared with the competitive methods such as e-beam lithography. In addition, the higher fabrication tolerance makes it an outstanding candidate for future application in photovoltaic and sensors. Not only MMPAs in visible frequency range, but those for other ones such as GHz and THz also need simpler design, easier fabrication, lower production cost, and higher fabrication tolerance [5].

As we move upward in the design stack to more complex components and devices, new emergent phenomena and more complex functionalities might be synthesized at each level, paving the way to a new era for MMPAs [8]. We feel confident that the MM revolution in the field of absorption has just started, and will present exciting new developments in the realm of practical applications in the coming years by overpassing the entry barrier and replacing the conventional applications using existing materials/structures, and by exploiting new ones.

An important potential application of MMs, which is relevant to MMPAs, is invisible uniforms/cloaks for soldiers and so on. Since J. Pendry in 2006 [9], many devices trumpeted as invisibility cloaks have been described, but they only work in the lab with specific wavelengths or from certain angles. The US Army has made a call for proposals in 2015 from companies for wearable camouflage with a chameleon-like ability to change according to the background. MMs are probably the best solution. Previous efforts in this field using technology like LEDs were hampered by power and computing requirements.

On the other hand, although they can bend light, perfect invisibility of macroscopic objects for all visible colors is fundamentally impossible. Bending light over the entire spectrum is forbidden by relativity. The wearer would be effectively transparent at some wavelengths but not all, rendering them as a colored shadow or ghost image. US Army need to work in all terrain from all angles. They also need to function across a wide range of temperatures, in rain and snow, and without hampering a soldier's normal duties. If the adaptive camouflage requires a power source, this must weigh no more than 0.45 kg and provide at least 8 h of operation. Every requirement looks challenging to the present researchers on MMs, but this might be possible in the future if we work hard and effectively.

Not only invisible uniforms/cloaks, but stealth science and technology based on microwave (L, X and K_u bands) MMPAs are going to be further developed. The objects comprise military aircrafts, missiles, ships, tanks and vehicles. The defense, the shielding and the cloaking of important military places and facilities such as communication command, based on MMPAs, are also investigated.

Ongoing and perspective researches on MMs comprising MMPAs can be mentioned to be, first of all, MM-based absorbing and radiating structures, which are directly relevant to the scope of book. This should cover all topics related to the use of MMs and metasurfaces in absorbing and radiating structures. Multi-band MMPAs using single pattern are interesting for the simple fabrication and the relevant advantages. In addition, bottom-up approach towards MMs and plasmonics should be developed. This covers all the aspects of novel approaches to manufacturing of materials with special EM properties as MMs and plasmonic materials. As next, structured light in MMs is going to be investigated. The synergy between MMs and structured light opens entirely new opportunities fundamental and applied optical science. Nonlinear and reconfigurable MMs and plasmonics are also promising. Nonlinear and reconfigurable MM and plasmonic devices are considered among the most promising platforms for efficient light manipulation and the generation of tunable functionalities in devices such as lasers, optical antennas, waveguides, switches, and modulators. Due to the favorable power-law scaling of near-field enhancements, new nonlinear optical properties are emerging in chiral meta-surfaces and MMs as well. Ultrahigh frequency transport and THz plasmonics in nano-scale structures, MMs, and 2-dimensional materials are also important. Ultrahigh frequency transport at time scale comparable or smaller than momentum and energy relaxation time becomes very important, since the mainstream semiconductor technology reached dimensions below 15 nm (smaller than the Si mean-free path at room temperature). Potential applications include ultra-high-speed wireless communications, imaging, sensing, and medical and biotechnology applications.

Functional meta-structures and nanomaterials: properties, fabrication and modeling is also studied. The field is focused on optically-active hybrid nanomaterials composed of semiconductors, metals, dielectrics, polymers, biomolecules, etc. Acoustic, elastic and thermal MMs are one of the most promising areas of wave propagation in complex media. Contrary to plasmonics, mechanical waves do not suffer from high metallic losses that prevent the applicability of many exciting concepts, such as negative refraction or invisibility cloaks. Next is parity-time (PT) symmetry in photonics, MMs and plasmonic systems. Recently considerable interests have been manifested for the novel physics and the novel opportunities offered by PT-symmetric systems that combine gain and loss in the same structure. Optical forces and manipulation of momentum in MMs and plasmonics should be mentioned. Indeed, it has been shown that MMs and plasmonic devices offer the possibility to enhance the optical momentum transfer and increase the associated optical forces by several orders of magnitude.

Finally, large-scale MM assemblies should be overcome. A pragmatic, nontrivial issue impeding the realization of optical MM devices is the need for robust and efficient assembly strategies. Industrial applications of MMs will be accelerated, which is the transition of MMs from basic and applied research into real world applications. The existing or planned MM components, devices or systems which are having an impact in any branch of science, engineering or technology development are included.

References

1. J.W. Park, P.V. Tuong, J.Y. Rhee, K.W. Kim, W.H. Jang, E.H. Choi, L.Y. Chen, Y.P. Lee, Multi-band metamaterial absorber based on the arrangement of donut-type resonators. Opt. Express. **21**, 9691 (2013)
2. P.V. Tuong, J.W. Park, J.Y. Rhee, K.W. Kim, H. Cheong, W.H. Jang, Y.P. Lee, Symmetric metamaterials based on flower-shaped structure. Mater. Chem. Phys. **141**, 535 (2013)
3. P.V. Tuong, Y.J. Yoo, J.W. Park, Y.J. Kim, K.W. Kim, Y.H. Kim, H. Cheong, L.Y. Chen, Y. P. Lee, Multiplasmon-induced perfect absorption at the third resonance in metamaterials, J. Opt. **17**, 125101 (2015)
4. Y.J. Yoo, H.Y. Zheng, Y.J. Kim, J.Y. Rhee, J.H. Kang, K.W. Kim, H. Cheong, Y.H. Kim, Y. P. Lee, Flexible and elastic metamaterial absorber for low frequency, based on small-size unit cell. Appl. Phys. Lett. **105**, 041902 (2014)
5. J.Y. Rhee, Y.J. Yoo, K.W. Kim, Y.J. Kim, Y.P. Lee, Metamaterial-based perfect absorbers. J. Electromagnet. Wave. **28**, 1541 (2014)
6. S. Borri, P. Patimisco, A. Sampaolo, H.E. Beere, D.A. Ritchie, M.S. Vitiello, G. Scamarcio, Terahertz quartz enhanced photo-acoustic sensor. Appl. Phys. Lett. **103**, 02110 (2013)

7. M. Hedayati, F. Faupel, M. Elbahri, Tunable broadband plasmonic perfect absorber at visible frequency. Appl. Phys. A **109**, 769 (2012)
8. F. Monticone, A. Alù, Metamaterials and plasmonics: from nanoparticles to nanoantenna arrays, metasurfaces, and metamaterials. Chin. Phys. B **23**, 047809 (2014)
9. D. Schurig, J.J. Mock, B.J. Justice, S.A. Cummer, J.B. Pendry, A.F. Starr, D.R. Smith, Metamaterial electromagnetic cloak at microwave frequencies. Science **314**, 977 (2006)

Index

A

Absorption-tunable MMPA, 1, 2, 5, 76, 92, 134, 136

Applications, 1, 3, 5, 28, 60, 69, 70, 73, 75, 93, 113, 144, 154, 158, 169, 171, 172

Asymmetry resonators, 106

B

Broadband, 1, 2, 5, 20–22, 24, 26, 29, 55, 56, 63, 64, 69, 77, 80, 84, 85, 87, 92, 95, 106, 110, 113, 122, 127, 131, 140, 154

Broadband absorption, 20, 21, 64, 87, 106, 122, 154

Broadband MMPA, 94, 138

C

Chiral metamaterial, 156

Circuit-analog absorbers, 20

E

Easier fabrication, 170

Effective-medium approximation, 23, 35

Electromagnetically-induced absorption, 101, 102

Electromagnetically-induced transparency, 1, 4, 99

Equivalent-circuit theory, 30, 39, 42

F

Finite-difference time-domain method, 45

Finite-element method, 45, 48, 83

Flexible metamaterial, 4, 64

Flexible MMPA, 69, 93, 95, 170

Frequency-tunable MMPA, 78

G

Geometrical-transition absorbers, 20

H

Higher/lower frequency, 61, 169

I

Incident-angle independence, 56, 71, 156

Industrial applications, 1, 5, 172

L

Localized surface-plasmon polariton, 55, 87

Low-density absorbers, 20, 24, 25

Lower cost, 170

M

Maxwell's equations, 11, 12, 39, 45, 46, 70

Metamaterial, 1, 20, 55, 89, 113, 143

Metamaterial-based perfect absorbers (MMPAs), 2, 3, 31, 32, 42, 55

MMPA operating in the GHz range, 55

MMPA operating in the IR/visible range, 79, 93

MMPA operating in the THz range, 69, 75, 119

Multi-band MMPA, 2, 5, 171

N

Numerical techniques, 45

O

Omni-directional MMPA, 77, 95

P

Perfect absorber, 1–3, 20, 55, 59, 64, 82, 102, 113, 143, 154, 169

Perfect electromagnetic-wave absorbers, 20

Perfectly-matched layer, 143, 160

Polarization independence, 3, 85, 144

Polarization-insensitive MMPA, 85

Y.P. Lee et al., *Metamaterials for Perfect Absorption*,
Springer Series in Materials Science 236, DOI 10.1007/978-981-10-0105-5

S
Salisbury screens, 20, 27, 29, 30
Simpler design, 170

T
THz gap, 69, 71
Transfer-matrix method, 45, 49
Transmission-line theory, 28, 39, 43

U
Ultra-broadband MMPA, 113
Uniaxial perfectly-matched layer, 144, 160

W
Wide-incident-angle MMPA, 5, 143, 165
Wider spectral regime, 169

The manufacturer's authorised representative in the EU is Springer Nature Customer Service Centre GmbH, Europaplatz 3, 69115 Heidelberg, Germany. If you have any concerns regarding our products, please contact ProductSafety@springernature.com

Printed and bound by CPI Group (UK) Ltd, Croydon, CR0 4YY
25/09/2026
02222850-0008